Project Control by Critical Path Analysis
a basic guide to CPA by bar chart

PROJECT CONTROL BY CRITICAL PATH ANALYSIS

a basic guide to CPA by bar chart

C. W. Lowe

BUSINESS BOOKS

First published 1979

ISBN 0 220 67012 9

Printed in England by T. J. Press (Padstow) Ltd.,
Cornwall, and bound by R. Hartnoll Ltd., Bodmin,
Cornwall, for the publishers, Business Books Ltd.,
24 Highbury Crescent, London N5.

Contents

Foreword

by John A. Rogers, PhD, MA, DIC, CEng, FIProdE, FIWM
*Head of the Department of Mechanical and Production Engineering,
Southall College of Technology*

In the early 1950s I visited my parents in North Wales while on leave from the Army. At that time my father was Plant Supervisor in the Ruabon Works of Monsanto Chemicals Ltd. He described to me a method of project control which was being introduced into the plant by its new manager, C. W. Lowe. I was highly impressed with this charting method whereby the shortest possible time for completing work could be easily calculated, together with associated resources and costs. Indeed, I soon realised that the method would stay in my mind forever.

On returning to industry I used the technique with success on a small scale during the late 1950s. However, my interest in the subject was further stimulated in 1961 whilst following a postgraduate programme at Imperial College. I had been given a GEC Scholarship to attend the course and met the title 'Critical Path Analysis' during my studies. Credit was being given to the Americans for developing the technique, but to me it was Lowe's method by another name though developed independently in America. Some of the problems which met me on my return to GEC as Production Controller of the Erith Works were 'crying out' to be solved by critical path analysis. The method was introduced to the works in 1962 and the first 'thousand-activity diagram' drawn in that year for the management control of the manufacture of a steam turbine.

The method was highly successful and was used on a wide range of projects during the remainder of my industrial career, first as Production Controller and later as Production Manager.

In 1966 I entered full-time education as a lecturer and came across Lowe's first book *Critical Path Analysis by Bar Chart*, which had just been published. Its publication was timely as the subject was now popular and finding its way into education syllabi. Existing books were 'heavy' and based on the work of the Americans. Lowe's book was well written, simple to understand and crammed full of examples based on his practical experience. I continued to use critical path analysis as a consultant in industry, with the new book 'at my elbow'.

Research carried out by Lowe and myself, working together, showed many advantages of his method over those developed later in America. It was therefore no surprise to me that the book became so successful in industry and education that it sold all over the world, including Japan. Further success followed and this book, *Project Control by Critical Path Analysis*, stems from the two previous editions of the first book.

I have lived with Lowe's technique for a quarter of a century and can think of no better one for project control. This book is the key to the technique and should be used by persons concerned with project management.

Acknowledgements

I thank Monsanto Chemicals Limited for their permission to publish the examples used in Chapters 2, 3, 4, 5, 6 and 7, for projects 1 to 10 in Chapters 8 and 9, and for projects 11 to 14 in Chapter 10. In particular I would like to express my thanks and gratitude to the works' engineers who have supplied the job sheets and job charts.

My thanks are also due to: Wates Limited for Project 20 and for the information contained in Figures 11.11 and 11.12; and Star Paper Mills Limited for Project 21 and for the information contained in Figures 11.13 and 11.14.

I would also like to thank my friends who have supplied data for all the other projects and example projects mentioned in Chapters 11 and 13, and for advice and discussion. Any reader who disagrees with the job sequence for any project in this book may have a more suitable sequence for his particular application, but the job sheets as given in Chapters 9, 10 and 11 demonstrate the enormous scope of job charts.

The job sheets quoted in this book are exactly as produced by the engineers concerned; no attempt has been made to 'straighten them out' in any way. The projects date from the late 1950s to 1976 so the job costs are in no way comparable; each job is the direct cost of completing the job at the time the job was completed.

Any opinions expressed in this book are my own alone.

C. W. LOWE

Chapter 1

Introduction

The efficiency of the engineering team in all aspects of maintenance, installation, and replacement is one of the factors which controls the profitability of an organisation. Time loss must be minimised. Sometimes a maintenance job is undertaken by one man and usually this man prepares a work schedule which may be written on job breakdown sheets or may be developed mentally; but however it is prepared the schedule enables the craftsman to work through all the steps of a job so that it may be completed with a minimum of hindrance and delay.

Mostly any item of maintenance, installation, or replacement consists of a number of distinct pieces of work and requires the time and attention of a number of men of different trades or skills. Any of these may be required to work at the same time on the same job or on different jobs, or may be required to work at different times on the same job or on different jobs. In cases like this we need some sort of procedure — a more sophisticated breakdown sheet — which will help us to ensure that all men and items of equipment are in the right place at the right time. Bar charts have been used for a long time to help reduce time losses, and the procedures of critical path analysis have been developed to provide the assistance necessary to minimise hindrance and delay. In this book bar charts will be used for critical path analysis.

First, four questions need to be discussed: Why do we need critical path analysis? What information is needed as input to critical path analysis? What information will be provided by critical path analysis? and Who is to perform the critical path analysis?

Why do we need critical path analysis? To start, *Project* is defined as the total work that has to be done from start to finish and *Job* as any distinct piece of work which marks a stage of the project. Jobs are not constant in time content or in work content; sometimes there is a restraint or restriction on the progress of the project which has time content but no work content and which must be considered as a 'job', and sometimes a score or more items listed as separate jobs in one project may be bulked together as one job in another project.

In any project the flow of work may be dependent on delivery of equipment or spare parts, or on the availability of labour, or on some preparatory work which itself may be dependent on delivery of materials. There may be a lot of other factors involved, and the more there are, the more can go wrong. It is not unknown for maintenance and installation projects to become somewhat confused because

1   Some item of equipment has not arrived at the promised time.
2   Non-arrival of some parts or material has delayed fabrication of other parts.
3   Completion of some job in the project has been delayed, possibly because of withdrawal of labour to another job on the same or a different project without the necessary attention having been paid to the comparative urgency of the jobs or the work content of the jobs.
4   Unforeseen circumstances, such as failure of a part or tool during the project, resulting in additional unscheduled work to be done.

Any one such occurrence might delay completion of a project, or at the very least cause delay to some job or jobs in the project, and result in craftsmen being idle until alternative work is found for them. But with sufficient forethought delays such as these can be minimised even if they cannot be eradicated altogether, and the techniques of critical path analysis will aid this minimisation. The procedures developed in critical path analysis can be of great assistance in the initial discussion and thinking stage of a project where the work is planned and scheduled, where the difficulties that are likely to arise are discussed and eliminated in advance, where provision is made for all the requirements of the work. Job Charts provide the necessary means to perform the critical path analysis, and the Work Charts derived from the Job Charts provide the means to control the progress of the work. The answer to the first question is that critical path analysis is needed to avoid getting into a mess.

What information is needed as input to critical path analysis? When it has been agreed to go ahead with a project, the first duty to be undertaken by the engineering staff is to prepare a plan of all the work to be done. The best plan for tackling any project can be devised only after careful consideration has been given to three items

1   *The technological sequence*   It is necessary to establish the technological sequence of the work involved. That is, it is necessary to determine the relationship of each job to all the other jobs because it is only by determining these job relationships that any planning can be done. This means that for every piece of work that we consider as a distinct job we must determine
    a   Which job must precede it, i.e. which job must be finished before the given job can be started.
    b   Which jobs may be carried out concurrently with the given job, and
    c   Which jobs must follow it, i.e. which jobs cannot be started until the given job is finished.

For example, in a project we might find that Job 8 must be completed before Job 9 can be started — we cannot put the gas taps on a bench until the bench is available; Job 10 can be carried on concurrently with Job 8

— installing the gas mains can go on at the same time as installing the bench; both Jobs 9 and 10 must be completed before Job 11 can be started — final connections between gas main and taps.

2   *Job durations*   It is necessary to establish the time requirement of each individual job in the project because it is only from these job durations that the overall duration of the project can be determined. For many projects a required date of completion may be stated before the planning of the project is started, and planning will show if this date is feasible. It may be necessary to estimate the likely effect on the operation schedule of a project completion date earlier or later than that first postulated as 'required'. And it may be necessary to reduce some job durations so that the required date of completion can be met.

3   *Alternative plans*   It is necessary to develop any alternative plans that may be feasible or that may be desirable. It is essential to consider if interruptions to any job are possible or even likely. Consideration must be given to the way in which any interruption on one job may affect completion of any other job, and the effect of any interruption on the project completion date must be determined. When the data derived under items 1 and 2 have been tabulated, the techniques of critical path analysis can be used to produce the initial plan and any alternative plans that may be required.

The answer to the second question is that input to critical path analysis consists of data about job sequence, job duration, and their alternatives.

What information will be provided by critical path analysis? Critical path analysis permits the determination of that sequence of jobs which starts right at the beginning of the project, which ends at the finish of the project, and for which there is no overlap of job times or free time between jobs. Each job in this sequence must be completed before the next can be started. This sequence of jobs is called the *critical path* of the project, and the jobs forming the critical path are known as critical jobs. The critical path determines the duration of the project, and this is the sum of durations of the critical jobs. The jobs on the critical path are those which must be controlled to prevent undue extension of the project duration and which must be speeded to permit completion of the project in a shorter time than that indicated at first. Every project must have a critical path; complicated projects may indulge in several critical paths or consist entirely of critical jobs. The jobs which are not on a critical path, the non-critical jobs, are those whose positions in time may be manipulated: for necessity, for critical jobs, or for convenience with respect to each other; and when the critical path has been determined, it is possible to fix time limits for the non-critical jobs. An earliest possible starting time and a latest permissible finishing time can be determined for these non-critical jobs, and for these jobs the time interval between these calculated times will be greater than the corresponding job durations. Critical path analysis consists of three phases which are interdependent but which may be considered separately for convenience. These phases are as follows:

1   *Planning* entails consideration of all the separate jobs in a project, determining the work content of every job, estimating the duration or time content of every job, and determining the interdependence and sequence of all the jobs. The output from this phase consists of a Job

Sheet and a Job Chart which are the input to the second phase.
2   *Analysing and scheduling* follow from the Job Chart which is the first feasible plan for the project based on the assumption that each job can be started at its earliest possible start point. The critical path can be identified and the overall project duration can be determined. The value of crashing any job or sequence of jobs can be estimated. Manpower and other resources can be scheduled and the feasibility of completing the project to standard time, minimum time or maximum profitability examined. A starting date can be set and a final plan agreed; this final plan is set out as a modified Job Chart called a Work Chart which also indicates resource requirements. The output from this phase is a Work Chart and, if required, a tabulation of work instructions, and these are the input to the third phase.
3   *Controlling* begins as soon as a feasible plan has been achieved. The Work Chart is used to direct and control all effort so that the latest permissible job finishing times are met and the project is completed as planned. The output from this phase is the completed project.
The answer to the third question is that the information provided by critical path analysis consists of:
a   A feasible plan for the project in which are determined the earliest starting time for each job and the latest possible starting time for each job that will permit completion of the project in the shortest possible duration.
b   A list of those jobs for which the earliest and latest permissible starting times are the same and which are the critical jobs.
c   A knowledge of exactly when each resource is required.
  Who is to perform the critical path analysis? The use of networks to describe the relationships between the jobs in a project has led to the development of somewhat complicated techniques for determining the necessary timing of the jobs, and the use of computers has been advocated to perform the calculations. Computers are wonderful time and labour saving machines that are of great value in many applications. They are used for the solution of complicated mathematical problems in scientific and engineering research, design and development where many hours of manual calculation can be saved, and they are used for repetitive accounting work and information retrieval where both time and staff can be saved. The purpose of a computer may be expressed simply as providing a means to increase man's productivity, and having said that two points should be made. Firstly, like any other item of equipment a computer should pay its way and should be used only when it is economic to do so. The saving resulting from the use of a computer may be expressed in terms of time or money, but if a computer does not produce a saving it is no better than any other status symbol. When additional human effort is employed to provide data for a computer the whole purpose of the computer is negated. Secondly, the existence of computers has led them to be used for jobs for which simpler and better methods are available, and the particular example with which we are concerned is the use of networks and computers for planning and scheduling a project.
  In recent years much has been written about the techniques of critical path analysis, but these techniques seem to have been neglected by industry except for a very few specialists. This may be because engineers and supervisors do

not understand networks, or do not understand or trust computers, or just do not want to be bothered with a new technique. Whatever may be the reason, it is a pity, because the techniques of critical path analysis are most useful and have much to offer everyone concerned with the planning, scheduling and controlling of maintenance, installation and replacement work. Some of the techniques in current use are too difficult and complicated to get over to the men who have to supervise and control the work, so that control of the work as it is carried out has become divorced from the planning and scheduling of the jobs in a project. This may or may not be good, but usually it means that the foreman in charge of a piece of work has no idea of the relative importance of that particular piece of work compared with any other piece of work in the project. In some projects, such as the building of a new generating station or a new factory, which involve many thousands of jobs needing the labour of many thousands of men over a long period of time, it is obvious that planning the jobs and scheduling the resources and controlling the work are quite distinct areas, and the use of computer procedures is essential to keep a continuing check on progress. But for every large project of that type there are tens of thousands of small projects relating to overhaul, repair, and minor installations, each of which has to be planned, scheduled, and controlled by one man or a small group of men working together. And the answer to the fourth question is that the critical path analysis should be performed by someone connected with the project — engineer, planner or foreman.

The definitions used in critical path analysis are as follows:

Project The total work that has to be done from start to finish. Usually a project starts at the time a request is made for work to be carried out. It may include such items as 'engineering evaluation', 'chemical research', 'engineering design', 'management approval', etc., which must be completed before any other work can be started.

Job Any separate piece or item of work which marks a stage in the project or makes a definite advance towards completion of the project. It is necessary to include each restraint such as 'waiting for equipment from suppliers', 'waiting for furnace to cool', 'purging and testing atmosphere' as a job; each may require some time even if it requires no effort on the part of any member of the engineering staff. But note that the definition of what constitutes a 'job' may depend on the project, the plan, or the schedule; in a small specialised project fitting a bolt to an inspection plate on the manlid of a vessel may constitute a job, while on another project installation of the complete vessel assembly may be listed as a single job. Jobs may be grouped or subdivided wherever feasible to meet time or resource restrictions.

Crash job A job whose duration can be reduced. Usually this reduction involves extra effort — more men on the job or overtime working — and this involves premium expenditure. Before agreeing to crash a job it is necessary to ensure that it is worthwhile incurring this additional expenditure.

Critical path That schedule which identifies the jobs that must be completed in succession, the starting and finishing time of each of these jobs, so that the project may be completed without delay. Any other path through the project is known as a 'slack path'.

Critical job   Any job on the critical path. The latest finish of one critical job must be the same as the earliest start of the next critical job. There is no free time or overlap between successive critical jobs.

Float   The amount of time that the start of a job may be delayed without affecting any other job. The first estimate of float is made on the assumption that all jobs are started at their earliest possible start times. If a sequence of jobs forms a simple chain the float is available to any job in the chain. There is no need to distinguish between free float, total float, independent float, or any other terminological float.

Floater   Any job whose position in time can be moved without affecting a critical job. Floaters are found on slack paths.

Job sheet   A list of all the jobs and restraints in the project giving all relevant sequence, duration and cost data for each job. Data may be included for standard working and for crash working.

Job chart   A diagrammatic representation of a project in which the job elements are drawn as parallel lines to a convenient time scale. There is no limit to the size of a job chart, which fulfils a dual purpose: it describes a collection of related jobs as a network, and at the same time it defines the time limitations of these jobs in a time chart. No mathematics are involved; no artistry is required; no special equipment is needed.

Job line   A line in the job chart which stands for a particular job in the project. Each job line is numbered to correspond with the number given to its job in the job sheet. Usually the job lines are drawn in vertical array.

Time line   A horizontal line in the job chart drawn at some particular point in time to relate the completion of two or more jobs to the start of another job or jobs. The first time line is called the 'start line' and the last time line is the 'finish line'. Where a time line is drawn to divide the job chart into distinct periods of time different time scales may be used in each period, but the scales must be made clear on the chart.

Now there are two items that should be mentioned briefly: bar charts and networks. Bar charts have been used for a very long time to help reduce time losses and to assist scheduling of labour and other resources, and the bar charts are easy to construct and easy to understand. A bar chart for a project consists of a number of horizontal lines or bars, each one representing a job and lying between some time limits. The one drawback of a bar chart is that it does not make clear any relationships that may exist between jobs, and we may have a bar chart for a project, from which the engineers are working, in which related jobs are widely separated by an array of other jobs so that it is virtually impossible to determine any relationships between the jobs. A network is a collection of connected lines, each of which indicates the movement of some quantity between two locations. Many different forms of network are in use in industry, perhaps under other names, in such fields as organisation and methods, work study, computer programming, and process and engineering studies; and some of the names that come to mind are block diagrams, flow charts and string diagrams. The quantities involved in these networks may be costs, distances, times, volumes, weights, or any other additive measure. Any collection of operations which have some dependence on each other can be described in network form. An arrow network is a diagram constructed to show how all the jobs in a project are related to each other and to the various stages of the project. An arrow network is constructed from the data provided in the job sheet and, except for very simple projects, is not easy to construct. In an arrow network the arrows are not drawn to scale, so that even if constructed successfully, the arrow network is of little use since it contains no information that has not been provided in the job sheet. The arrow network is a diagrammatic representation of the project and has to be worked on to provide any further information; and this work can be done from the job sheet. Nye Bevan may have had arrow networks in mind when he used the phrase: 'All facts and no vision'. Whether he did or not, job charts are proposed for critical path analysis to represent the facts and present the vision. The job chart is a combination of a bar chart and a network, or, if preferred, a bar chart drawn in network form so that it has all the advantages of a bar chart in showing when the jobs must be done, all the advantages of a network in showing the relationships between jobs, and none of the disadvantages of arrow networks in that there are no difficulties in

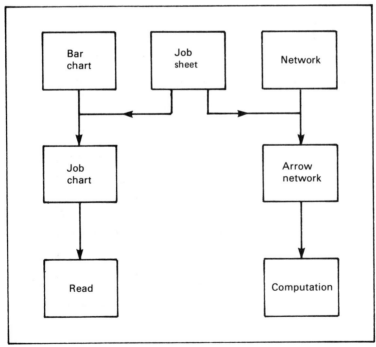

Figure 1.1 Analysis of a project

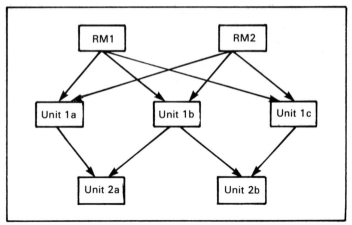

Figure 1.2 Part of a process network

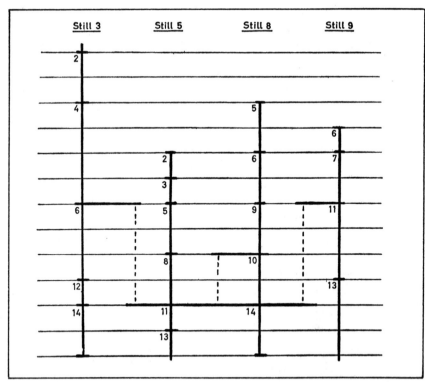

Figure 1.3 Fraction blending chart

construction and no mathematical work to be performed on the completed chart.

Figure 1.1 shows the relationship between the arrow network and the job chart, but it must be emphasised that the arrow network gives exactly the same information that is on the job sheet, no more and no less, whereas the job chart gives all the information that is available about the project and about each individual job.

Plant chemists and engineers have used time charts ever since the first of them wanted to show what happens when units run out of phase, and networks, of which the example shown in Figure 1.2 is a small part, are commonplace in the chemical industry. My first real use of process time charts arose in 1950 when dealing with the refining of tar acids. At the Ruabon Works of Monsanto Chemicals Limited there were then ten stills which refined and fractionated several different tar acids and their associated fractions. It was necessary to estimate, with reasonable accuracy, just when the required fractions would become available for blending or for refractionation, and it seemed easiest to make this estimate on a time chart, which consisted of a vertical array of fraction lines, drawn to time scale, with horizontal time lines showing when the necessary fractions would become available for blending. A greatly simplified fraction chart might be as shown in Figure 1.3, which shows the estimated time of availability of fractions 4, 7, 8 and 9 for making a blend of these fractions.

There was no difficulty in transforming the fraction charts with their fraction lines into job charts with their job lines. The only difference between them is that the fraction chart presents an array of the estimated times when fractions from different stills will become available, giving the times of the start and finish of distillation of each fraction; while the job chart presents an array of the estimated times when jobs in the same project will be completed, giving the estimated start and finish time for each job.

Chapter 2

The Job Sheet

Before any planning or scheduling can be started on any project it is necessary to know exactly what work is involved. In order to obtain this knowledge the total work that has to be done must be divided into its elements, each of which we call a 'job'. Each of the jobs in a project must permit discussion and evaluation of itself and by itself, whether or not it may be carried out more conveniently with some other jobs. Each job must stand as a distinct entity. The amount of work to be done in a 'job' is not constant; the labour employed and the time required for completion will vary with the idea of what constitutes a job. The requirement is that a job must be an item or piece of work which marks a stage in the project or makes a definite advance towards completion of the project, and which may be considered separately from all other jobs. This separateness must be emphasised; confusion has arisen when two jobs have been allocated an overlap in work content.

It may be convenient to aggregate a lot of very small job-elements into one job as, for example, we might decide to have

Fitting gas-taps to benches

as one job rather than have forty or fifty jobs each dealing with the fitting of a gas tap in its particular location. This is permissible as long as *all the work* involved in fitting *all the gas taps* can be carried out in sequence unaffected and uninterrupted by any other jobs. At other times it might be convenient to divide a job into smaller sections, each of which can be dealt with separately as, for example, with the job

Install new water distribution main

which might be divided into jobs according to physical layout, starting with

Install new water distribution main from Pumphouse to Building A

followed some time later by

Extend new water distribution main from Building A to Building B, etc.

Such subdivisions may be made whenever it is easier for the project planner to consider the smaller jobs, perhaps thinking ahead to the scheduling of resources which will have to be made, than to consider the aggregate job. As well as the items of work it is necessary to include as a 'job' any item which may cause some restriction to the smooth flow of work — a restraint — which must be overcome before work can continue. A restraint is a time-consuming element of the project, which may or may not have a work content, if it has some work content, generally this will be outside the control of the project supervisory staff. Restraints may include such items as

> Waiting authorisation by management
> Job allocation and planning, or
> Waiting delivery of equipment, etc.,

before work can be started on site; or such items as

> Chemical atmosphere test,
> Equipment cleaning before removal, or
> Waiting for furnace to cool before entry, etc.,

which are met during the course of the work. All such items must be included in consideration of the project.

The first step in the planning of any project is to compile a job sheet. This is a list of all the jobs and restraints known to be involved in the project and is prepared by, or with the assistance of, the project engineer. Further restraints may arise because of sequence relationships between jobs in the project, and these will be dealt with during the planning stage as they become known. The jobs are listed and numbered, not necessarily in the order in which they are to be tackled, though if this can be done it makes a very neat job sheet; if this order is required the job sheet can be rewritten, though in most cases this would be a waste of time. The first job on the sheet is numbered 1, and numbers for the other jobs follow in order down the sheet, regardless of the sequence of performance of the jobs. In projects which have 26 or less jobs the jobs may be identified by letters, but where there are more than 26 jobs the use of letters and letters with subscripts tends to be confusing and it is good practice always to number the jobs.

Next it is necessary to provide information regarding the relationships between jobs, and this is done by means of a sequence code written alongside each job in the job sheet. Each sequence code is written in two parts, separated by a colon or semicolon. For each job the first part of the sequence code states that stage of the project at which the job is to start, that is, which other jobs must be finished before the job concerned can be started, and *all* the precedent jobs required to finish at this stage must be included; the second part of the sequence code indicates the stage of the project reached when the job is finished, and the number used here is the job number. This arrangement of sequence codes has become standard practice, although strictly speaking the second part of the sequence code is superfluous as it repeats a number already written on the same line of the job sheet; the first part of the sequence code is the essential part and could be tabulated under some such title as 'Jobs to be Finished'. It has been found easier to look at a sequence code such as 15, 16, 17:29 and say to oneself 'Jobs 15, 16 and 17 must all be finished, when Job 29 can be started' than to look at the numbers 15, 16 and 17 and then swivel one's

eyes across the sheet to ascertain the number of the job to follow these three.
Examples of sequence codes of the type commonly used are:

Job 2: Job Sequence 0:2. The zero indicates that Job 2 may be started
right at the beginning of the project; stage zero is used to indicate the
start of the project, as at this point no work has been done.
Job 17: Job Sequence 4, 6, 11:17. This indicates that *all three* of jobs 4, 6
and 11 must be finished before Job 17 can be started.

A series of sequence codes such as:

4, 5:7   4, 5:8   4, 5:9

indicates that each of Jobs 7, 8 and 9 may be started and carried out
concurrently when *both* of Jobs 4 and 5 have been finished.

If all that is desired of the critical path plan is identification of the critical
jobs in the project, the only other information required on the job sheet is the
expected duration of each job. This is listed as *'standard time'*, and to avoid
any possibility of confusion the same unit of time should be used for every job
on a job sheet. If the expected durations of some jobs, for example, are 10
minutes, 15 minutes, and 45 minutes, while that of another job is 2 hours, this
latter duration should be entered as 120 minutes unless the scale representation used permits job durations to be entered as fractions of an hour.

The standard time is derived by the compiler of the job sheet using all
available knowledge and experience. If any particular job has been done on
previous occasions, the standard time estimate is likely to be accurate. Where
a job is to be done for the first time, knowledge and experience of the men, the
methods of work to be used, the conditions under which the work is to be
carried out, and the results of any similar work done in the past are brought to
bear, and the best possible estimate of the standard time is made. Every first-
time job automatically becomes the subject of careful supervision, which may
result in an improvement on planned performance and which provides
accurate data for any future repetition of the job.

Now maintenance of plant and installation of equipment are necessary
parts of the structure of industry and, like all the other parts, are required to
be controlled efficiently. The only valid measure of efficiency is the cost
incurred, and planning a project (and the scheduling which will be involved
some time later) can be improved if the costs of doing the work are
considered. So both 'standard time' and 'standard cost' should be entered on
the job sheet. At this time the standard cost of any job should be the total of
the direct costs involved in doing the work: materials, labour, and immediate
supervision. These are the costs which are under the control of the project
supervisor, and additional costs such as works indirect costs or overhead
charges should not be included in a standard cost; these indirect costs are
considered at a later stage of the project planning.

Further reflection may show that the job sheet includes some jobs which
could be completed in shorter times than those tabulated as 'standard' if
additional facilities could be made available and the consequent enhanced
costs of the jobs could be tolerated. These additional costs may be incurred
for such items as additional workmen or contractors on site either by
employing more men or by permitting overtime working, premium costs for
quicker delivery of materials from suppliers, hiring of additional machinery

or tools, etc. The estimated shorter time and total cost — standard cost plus premium charges — are entered in the job sheet as 'crash time' and 'crash cost'. If there are several possible crash times for any job, each with its associated crash cost, all should be entered and a separate column in the job sheet allowed for each. But note that every crash time entered must be feasible and capable of achievement; for instance, if the standard time for two men to complete a certain job is ten days, it is not permissible to quote a crash time of four days, assuming five men on the job, when it is known that only three men can be accommodated on the site. The inclusion of one non-feasible job time in a large project could wreck the schedule and wipe out any benefit the project should have received from planning.

Even for a small project with few jobs it is seldom possible to determine from inspection of the job sheet just which are the critical jobs and exactly what relationships exist between critical and non-critical jobs; with a project of more than twenty jobs such determination is impossible. Instances arise when non-critical jobs are rushed needlessly, to the detriment of critical jobs, and instances arise when non-critical jobs are left until they become critical and panic action is taken. There must be many occasions when critical jobs are not recognised as such and are neglected; usually such occurrences result in a state of utter confusion, so that completion of the project depends more on chance and personal goodwill than on probability and good intent. Thus it is necessary to transpose the information on the job sheet into a form that permits determination of the critical jobs and at the same time defines in detail the relationships between all the jobs in the project. The usual forms are diagrammatic and may be shown as an arrow network supplemented by a job matrix or some other method of counting through a network, or as a Job Chart. The job chart offers many advantages over the arrow network and I propose to spend very little time dealing with the construction of arrow networks, though in some of the examples discussed later we shall make a comparison between arrow networks and job charts, and you will be able to decide which is the easier to use and which offers the more advantages.

It has been stated that compilation of the job sheet can be facilitated if the diagram and job sheet are compiled together, job by job, and there is no reason why this should not be done whenever it is thought desirable. If joint compilation helps in any way then joint compilation should be used but it must be emphasised that all the work entailed in the project must be listed in the job sheet. In the example projects we shall discuss, we shall assume that the job sheets are completed first and are made available to the planners before any diagrammatic representation is started.

A small project which was divided into fourteen separate jobs was carried out at the Newport Works of Monsanto Chemicals Limited. We shall use this example for discussion of the various procedures involved. Figure 2.1 shows the relevant job sheet, from which the names of the individual jobs have been omitted for brevity. This job sheet contains all the data relating to the project except that it does not show clearly how the individual jobs are related to each other in time. All the work that is done on an arrow network can be done directly from the job sheet by adding two columns, for each of standard and crash if required, and labelling these 'float' and 'project duration'.

Figure 2.2 shows the float-project duration sheet for the project of Figure 2.1, and this sheet is compiled as follows:

Job number	Job sequence	Standard Time	Cost	Crash Time	Cost
			£		£
1	0;1	3 days	200	2 days	400
2	0;2	6	550	4	800
3	0;3	5	500	3	650
4	1;4	5	750	3	1,000
5	2;5	4	600	2	950
6	2;6	6	850	4	1,050
7	2;7	3	400	2	600
8	Dummy	—	—	—	—
9	3, 7;9	4	300	3	400
10	4, 5;10	8	1,100	6	1,400
11	4, 5;11	4	500	2	700
12	3, 6, 7, 11;12	6	650	4	900
13	3, 6, 7, 11;13	5	300	4	400
14	10, 12;14	3	250	2	350
15	9, 13;15	7	950	5	1,300
	Total	?	£7,900	?	?

Note: The query marks in the total columns indicate that these totals cannot be determined by straightforward addition of the column elements. The total sums can be obtained only by critical path analysis and are derived in the text.

Figure 2.1 Job sheet for plant repair

Job Sequence	Job Duration	Float	Project Duration
0: 1	3	a	3
0: 2	6		6
0: 3	5	4	5
1: 4	5	2a	8
2: 5	4		10
2: 6	6	2	12
2: 7	3	5	9
3,7: 9	4	6	13
4,5:10	8	2b	18
4,5:11	4		14
3,6,7,11:12	6	b	20
3,6,7,11:13	5		19
10,12:14	3	3b	23
9,13:15	7		26

Figure 2.2 Float-project duration compilation from Figure 2.1

14

1   Jobs 1, 2 and 3 can all be started immediately, and we can write the project durations on completion of these jobs as 3, 6 and 5 days respectively. For the time being we have no indication of any float attaching to these jobs.

2   Job 4 follows Job 1 and is complete in $5 + 3 = 8$ days. There is no indication of any float.

3   We see that each of Jobs 5, 6 and 7 follow Job 2 and are complete, respectively, in $4 + 6 = 10$ days, $6 + 6 = 12$ days, and $3 + 6 = 9$ days. We still have no indication of any float.

4   Jobs 3 and 7 must both be complete before Job 9 can be started. Job 7 has the later finish of these at 9 days and so Job 3 has $9 - 5 = 4$ days float. Job 9 is complete in $4 + 9 = 13$ days.

5   Both of Jobs 4 and 5 must be complete before Jobs 10 and 11 can be started. Job 5 has the later finish of these at 10 days and so Job 4 has $10 - 8 = 2$ days float. But this float follows the sequence Job 1 + Job 4 and may be shared between these two jobs, which we label 'a'. Job 10 is complete in $8 + 10 = 18$ days; Job 11 is complete in $4 + 10 = 14$ days.

6   All of Jobs 3, 6, 7 and 11 must be complete before Jobs 12 and 13 can be started. The latest finish of these is Job 11 at 14 days and so Job 6 has $14 - 12 = 2$ days float, and Job 7 has $14 - 9 = 5$ days float but since we have related Jobs 3 and 7 previously some of this float could be added to the float at Job 3. Job 12 is complete in $6 + 14 = 20$ days and Job 13 is complete in $5 + 14 = 19$ days.

7   Job 14 cannot be started until both Jobs 10 and 12 are complete. Job 12 has the later finish of these at 20 days, so that Job 10 has $20 - 18 = 2$ days float. Job 14 is complete in $5 + 14 = 19$ days.

8   Job 15 cannot be started until both Jobs 9 and 13 are complete. Job 13 has the later finish of these at 19 days so that Job 9 has $19 - 13 = 6$ days float. Job 15 is complete in $7 + 19 = 26$ days.

9   Therefore the overall project duration is 26 days — the biggest cumulator we have obtained. Since Job 14 is complete after 23 days it has $26 - 23 = 3$ days float, but this may be shared with Jobs 10 and 12, which have no other commitments, and these are labelled 'b'.

The critical path consists of those jobs which have no float and these are Jobs 2, 5, 11, 13 and 15; and we see that the sum of the durations of these jobs is $6 + 4 + 4 + 5 + 7 = 26$ days.

We have used the job sheet to calculate the project duration and to determine how much float is available to each of the non-critical jobs; but we cannot see how the various jobs are related in time, and it is advantageous to draw a job chart. The job chart shows the job relationship and also shows float and job and project durations without the laborious arithmetic necessary to deduce these from the job sheet.

Chapter 3

The Job Chart

Enough has been written and spoken to let it appear that critical path analysis of even a small project is beyond the reach and understanding of ordinary people and is sufficiently complex really to demand the use of a computer. But nothing could be more untrue. Determination of the critical path of jobs through a project, the relationships of jobs with each other, and the dependence of jobs on each other is simple, provided that the desire to introduce complications is rejected. Indeed, the determination may be made by any person who can draw straight lines to a predetermined scale. No mathematical expertise is required, since no mathematical procedures are involved in the preparation of the scale charts or in any analysis of the project that may be made from them. When analysis of a project requires more time and effort for manual solution than would be required for solution by computer, then a computer should be used; such instances arise when dealing with very large projects which extend over long periods of time and for which regular and frequent statements of the status of the project may be required. However, the great majority of projects met in industry are small and may be analysed quickly by the use of a simple manual technique which has proved to be completely successful in all applications to date.

The simple critical path plan has been called a *job chart*, and it not only forms the basis for the preparation of a work-plan for a project and provides the necessary assistance for scheduling resources, but also permits easy control of the progress of all the jobs in a project. Because each job is represented by a straight line drawn to scale — the job line — a job chart is drawn most easily on squared or ruled paper. Each job line is drawn to scale according to its duration and in sequence as directed by the job sheet, so that when using ruled paper the interval, or spacing, between rulings is taken to be whatever is the convenient unit of time. The advantage of using squared paper, especially that with ten or twelve divisions to the inch, is that smaller units of time may be used without inconvenience and the spacing between parallel job lines may be made uniform or varied to permit grouping of associated job lines; most of the examples described in later chapters have been drawn on ten-to-the-inch squared paper. Preparation of a critical path plan for a large project with many jobs is no more difficult than preparation of a critical path plan for a small project with few jobs if the plan is prepared in the form of a job chart; the larger project merely requires a larger sheet of paper and a proportionately longer time to draw the greater number of job lines.

Construction of a job chart begins with a horizontal line drawn across the top of the sheet. This line represents the start of the project, prior to which no work has been done and no time has been utilised and is taken as time zero. Usually this start-line is labelled 'Start', but as and when the need arises the start-line may be given a date and time so that the expected dates or times of completion of all the jobs in a project may be stated. From this start-line the job lines are drawn in vertical array in sequence as instructed by the job sheet, and are numbered so that each job line may be identified with its corresponding job in the job sheet. A short horizontal line is drawn to indicate the end of a job line, i.e. the finish of a job, and this permits separation of individual jobs in a sequence. A longer horizontal line is drawn on the chart to indicate time relationships between jobs; at any place in the project where two or more jobs must be completed so that some other job or jobs can be started, the horizontal line is drawn at the end of that job line which is latest in time, and is extended to connect all the jobs concerned. These horizontal lines are called time lines. And at any place in the project where two or more jobs may be started when a job has been completed, the time line again is extended to connect all the jobs concerned.

In the job chart each job line is drawn immediately beneath its appropriate starting time line, though there is nothing to prevent a time line being drawn right across the chart, and some of the succeeding jobs forming a separate array from that of the jobs preceding the time line, if such a procedure clarifies any points at issue. Construction of the chart is flexible. Each floater is followed by a dotted line which extends from the end of the job line to the next time line directed by the job sheet, and the length of the dotted line indicates the amount of float. There is no need for the inclusion of 'dummies', which merely appear as dotted lines between the appropriate time lines and are not numbered. The critical path through the project is indicated by those job lines which form a continuous line through the chart. Most often a chart is drawn to show the critical jobs 'stepped' on time lines, but if thought necessary or desirable the critical path can be drawn in the chart as a single straight line extending from start to finish and divided by time lines and job end lines. At the end of the last job line on the chart, the final time line is drawn across the chart and this is labelled 'Finish'; the interval between the start and finish time lines is the total time required of the critical path and, of course, is the overall duration of the project.

All the relationships which exist between the various jobs of the project may be seen at a glance from the job chart. All of the jobs which are concurrent are shown to be so, and all jobs in sequence are indicated clearly. The only difference between a scale arrow network and a job chart is that the arrow network insists that a point in time must be represented by a point in the network, while the job chart permits representation of a point in time by a horizontal time line of any length to suit the requirements of the chart construction and meet any personal preferences of the constructor. Whereas in an arrow network concurrent activities in a sequence all radiate from an event, so that after only two such events it is difficult to determine when an activity may be expected to start or finish, in the job chart all job lines are parallel, and relationships and start and finish times are immediately apparent. Jobs which may be carried out concurrently are immediately seen to be so, as they will be represented by parallel job lines drawn in the same interval of time.

On the job chart critical jobs, non-critical jobs which may become critical on reduction of a critical job duration or on extension of their own durations, and non-critical jobs which must remain non-critical are all shown clearly. Thus there is no need to attempt any arithmetical calculation from the job chart; in fact it is extremely difficult to imagine any calculations that could be made from the job chart even if one were obsessed with a desire for calculating. All job durations are indicated by the lengths of the job lines, so there is no need to write the job durations on the chart, and the start and finish time of every job can be read directly from the job chart. Since the job chart deals only with the jobs, there is no sequence of event numbers to write on the chart; the events of the arrow network are the time lines of the job chart. Every job chart is started at the top of the sheet, which means that the flow of work will be downwards in every case, so there is no need to direct the job lines; of course, the job chart may be drawn from left to right across the sheet if required, but the job lines must remain parallel, and in this case the flow of work would be from left to right. The job chart is a much simpler and less cluttered representation of the project than is an arrow network.

Figure 3.1 shows the job chart for the project detailed in the job sheet of Figure 2.1, and we now describe in detail the construction of the job chart according to the procedure given above. The start line is drawn across the top of an ordinary ruled sheet of paper and is labelled 'Start'; this is stage zero and time zero of the project. For the construction of this job chart it is convenient to take one line space to represent one day's duration so that the lengths of the job lines are measured in 'lines' on the sheet. Now the chart is constructed according to the instructions given under 'Job sequence' and 'Standard job time' in the job sheet. Each of Jobs 1, 2 and 3 may be started at the beginning of the project, so to correspond with these jobs, job lines 1, 2 and 3 are drawn respectively 3, 6 and 5 lines long from the start line, and a short horizontal job end-line is drawn at the end of each of these job lines. There is no particular method of arranging these job lines that can be recommended; it seems reasonable to start with job line 1 at the left-hand side of the chart, draw job line 2 in the middle, and draw job line 3 at the right-hand side. Left-handed people might reverse this order, but the actual positions of the job lines are not of real importance.

We see from the job sheet that Job 4 may be started as soon as Job 1 is finished, so job line 4 is drawn 5 lines long immediately after the end-line of job line 1, and is completed with a job end-line.

Next we see that each of Jobs 5, 6 and 7 may be started when Job 2 is complete, so the job end-line at the end of job line 2 is extended into a time line and job lines 5, 6 and 7, again ordered from the left, are drawn from this time line for 4, 6 and 3 lines long, respectively, and are completed with job end-

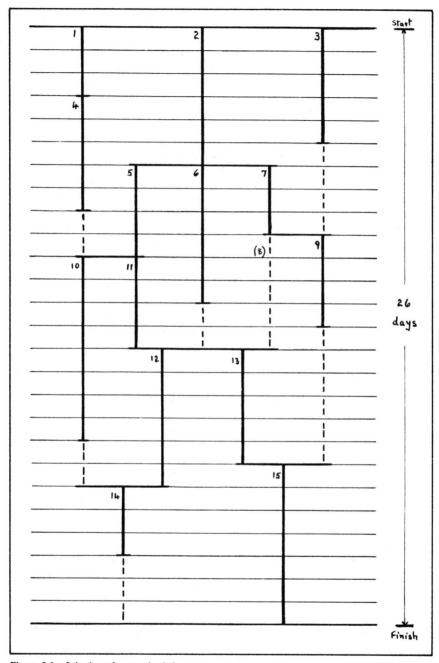

Figure 3.1 Job chart for standard times

lines. Job 8 was the job sheet number given to the dummy needed in the arrow network but it is not necessary to include this relationship as a distinctive job line in the job chart. The dotted line corresponding to the dummy has been labelled (8) in the job chart.

Now the job sheet indicates that Job 9 can be started only when both of Jobs 3 and 7 are complete and a time line must be drawn at the end of whichever of these job lines has the later finish. Here the time line is drawn at the end of job line 7, and a dotted line is drawn from the job end-line of job line 3 to this time line; this dotted line shows the amount of float available to Job 3 to be 4 days. From the time line job line 9 is drawn 4 lines long and is completed with a job end-line. The job sheet shows that both Jobs 10 and 11 may be started when both of Jobs 4 and 5 have been completed. Again a time line must be drawn at the end of whichever of Jobs 4 and 5 finishes the later, so here the time line is drawn at the end of job line 5 and a dotted line is drawn from the job end-line of job line 4 to this time line; this dotted line shows the amount of float available to the two-job sequence of Jobs 1 and 4 to be 2 days. From the time line, and ordered from the left, job lines 10 and 11 are drawn 8 and 4 lines long respectively and are completed with job end-lines.

Next, we see from the job sheet that each of Jobs 12 and 13 may be started only when all of Jobs 3, 6, 7 and 11 are complete. The latest finish of any one of these job lines is the finish of Job line 11, so a time line is drawn at the end of job line 11, and a dotted line is drawn from the job end-line of job line 6 to this time line; this dotted line shows the amount of float available to Job 6 to be 2 days. A time line has been drawn already relating job lines 3 and 7; now a dotted line is drawn from this time line to the time line at the end of job line 11, and this dotted line, as mentioned above and in the next chapter, corresponds to the dummy activity required in the arrow network. The time line at the end of job line 11 now relates the completion of Jobs 3, 6, 7 and 11 to the start of Jobs 12 and 13, and from this time line job lines 12 and 13, ordered from the left, are drawn 6 and 5 lines long, respectively, and are completed with job end-lines. Note that the dotted line (8) shows that Job 7, and Job 3 of course, has a five-day float with respect to the starts of Jobs 12 and 13; the value of this float to Jobs 3 and 7 is discussed later.

From the job sheet we see that Job 14 may be started only when both Jobs 10 and 12 are complete. Once more a time line is to be drawn at the end of whichever of these job lines finishes the later, and in this case the time line is drawn at the end of job line 12. A dotted line from the job end-line of job line 10 to this time line shows the amount of float available to Job 10 to be 2 days. From the time line, job line 14 is drawn 3 lines long and is completed with a job end-line. Lastly, the job sheet shows that Job 15 may be started only when both of Jobs 9 and 13 are complete, and a time line is to be drawn at the end of whichever of these jobs finishes the later. Here the time line is drawn at the end of job line 13, and a dotted line drawn from the job end-line of job line 9 to this time line shows the amount of float available to Job 9 to be 6 days. From the time line, job line 15 is drawn 7 lines long and is completed with a job end line. Now all job lines have been drawn on the job chart and the finishing time line is drawn across the sheet at the end of whichever job line finishes the latest; we have a choice from job lines 14 and 15 with job line 15 showing the later finish. Hence the finishing time line is drawn at the end of job line 15 and is labelled 'Finish', and a dotted line drawn from the job end line of job line 14 to this finishing line shows the amount of float available to Job 14 to be 3 days.

The above description of the construction of the job chart has taken quite a time to write; the job chart itself is drawn in less than a minute, the time requirement being that of reading the instructions in the job sheet and then drawing the job lines to the right lengths. Once the job chart has been drawn the critical path is immediately obvious; it consists of the continuous sequence of job lines through the job chart and is the sequence of job lines 2, 5, 11, 13, 15. All floats are shown clearly, and it may be seen immediately which jobs may be carried out concurrently. In fact, all the information required for scheduling and progressing the project is now available on the job sheet and job chart. The overall project duration, as indicated by this first feasible plan, is seen to be 26 days, controlled by Jobs 2, 5, 11, 13 and 15.

At this point it may be appropriate to consider the floats attaching to the various job lines shown in the job chart. As far as this first feasible plan is concerned, it is obvious that

Job 14 has 3 days float,
Job 6 has 2 days float, and
Job 3 has 4 days float,

and these floats are not affected by any other job. The two-job sequence, or chain, of Jobs 1 and 4 has a 2-day float, which may be utilised on either job or shared between them. In the case of the sequence containing Jobs 3, 7 and 9, the position is a little different. There is a five-day float after Job 7 and a six-day float after Job 9, but the Job 7 float and Job 9 start from the same time line and occupy the same interval of time, so that:

1   If Job 7 moves in its float, that is splits the float so that some precedes the job with the rest of the float following the job, the duration of the leading float must be added to Job 3 float and also subtracted from Job 9 float.

2   If the duration of Job 7 is extended, the float attached to Job 9 must be reduced by the amount of the extension; and conversely, if the duration of Job 7 is reduced, the float attached to Job 9 must be increased by the amount of the reduction.

Such items as these are seen immediately from the job chart but cannot be obtained from an arrow network without much calculation.

The relationship between bar charts, job charts and arrow networks

To digress for a moment or two, the differences and the relationships between bar charts, job charts, and arrow networks will be discussed.

Until a few years ago any maintenance or construction supervisor who had compiled the job sheet for the project, as in Figure 2.1, probably would have considered only the standard times of each job; and in order to obtain some idea of how the work should proceed, he would have constructed a bar chart something like that of Figure 3.2. If the supervisor had realised the significances of the job sequences the chart might have been re-arranged into the bar chart of Figure 3.3. This bar chart would have enabled him to schedule his manpower and maybe other resources and indicates when each particular resource is likely to be required.

The bar chart fulfils its purpose admirably. It is easy to construct and gives

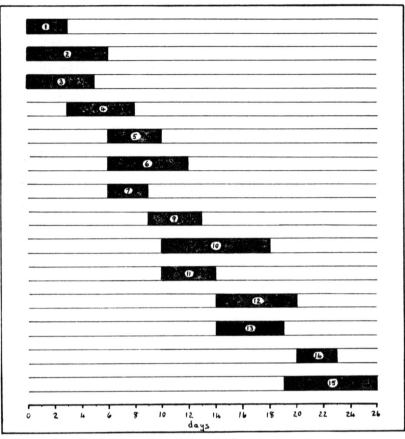

Figure 3.2   Bar chart for the project of Figure 2.1

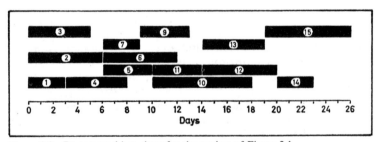

Figure 3.3   Re-arranged bar chart for the project of Figure 2.1

a visual representation of the whole project which shows exactly when each job is supposed to start and finish. The only information not provided on the bar chart is the sequence relationship between jobs, and if we add the necessary time lines and relationship dotted lines to the bar chart of Figure 3.3 we derive the job chart shown in Figure 3.4. In fact, the job chart is a simple adaptation of a bar chart which merely emphasises the job sequence and shows clearly the relationship between jobs. This job chart is identical in structure with the chart of Figure 3.1; it is drawn with job lines in horizontal array with the work flow from left to right instead of the vertical array with downward workflow as in Figure 3.1. The job chart provides some information additional to that given by the bar chart, and in the job chart we have a simple visual representation of all the information that is available about the project.

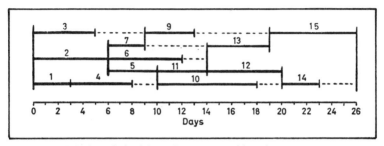

Figure 3.4 Job chart derived from the re-arranged bar chart

Now we can start imposing restrictions on our chart construction, and the first is that a point in time must be represented by a point in space. Of course, with this restriction in force, it is no longer possible to show all the jobs of a project as an array of parallel lines on a chart as the time lines are reduced to points. Representation of the project becomes that shown in Figure 3.5 in which each job line and its float remain in a straight line. With elimination of time lines it becomes necessary to introduce a dummy wherever it may be necessary to indicate a relationship between jobs in different sequences. In our example project it is necessary to introduce the dummy, 8, to indicate the dependence of Jobs 12 and 13 on Jobs 3 and 7. Because all job lines are not parallel it is considered necessary to direct the job lines so that the direction of the workflow is beyond doubt. At this point we have constructed an arrow

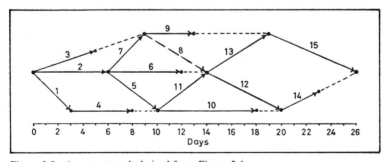

Figure 3.5 Arrow network derived from Figure 3.4

network in which the critical path may be picked out easily, as it is still the only continuous line through the project, and in which it is still possible to show job durations and floats to an agreed time scale. Even so, the network constructed in Figure 3.5 may be considered fairly reasonable, though it is much more difficult to construct than the job chart; whether or not it bears any resemblance to the original bar chart is a matter of personal opinion.

So we introduce the next complication by imposing the restriction that a job line and any float that may be associated with it shall not necessarily lie in one straight line. But note that this imposition is a restriction and not a relaxation, as an additional requirement has been imposed on the construction and the network has become more confused. This permits us to draw any number of arrow networks, such as Figure 3.6, each of which may be said to

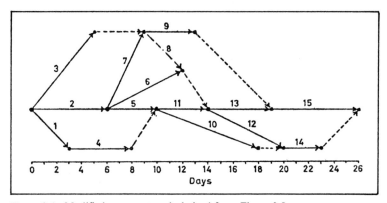

Figure 3.6 Modified arrow network derived from Figure 3.5

portray the same project, and three possible networks of this construction are shown in Figure 3.7.

Now it has become necessary to direct all the float lines. It is still possible to identify the critical path without difficulty as it remains the only continuous line through the project, and it is still possible to show job durations and floats to an agreed time scale. This far the lines of the network may be called job lines, and each line shows the duration and relationships of its job. There is no resemblance between this arrow network and the original bar chart, and from any of the networks of Figures 3.6 and 3.7 it is extremely difficult to determine which parts of which jobs and floats may be concurrent without drawing vertical lines across the network.

The next restriction to be imposed is that there shall be no representation of float whatsoever. This means that we must also do away with any attempt to represent a job by a job line drawn to scale, and it is no longer reasonable to use the term 'job line'. All jobs which must start from (or terminate at) a stage of the project are to be shown as arrows starting from (or terminating at) that stage, and in Figure 3.8 we have the familiar arrow network.

Now the arrows of the network merely indicate job sequence. A new terminology of *activities* and *events* is introduced. The network itself is meaningless, and in order to provide any useful information the job durations have to be marked on the appropriate arrows. It is not possible to identify the critical path by inspection, and there is no indication whether or not there is

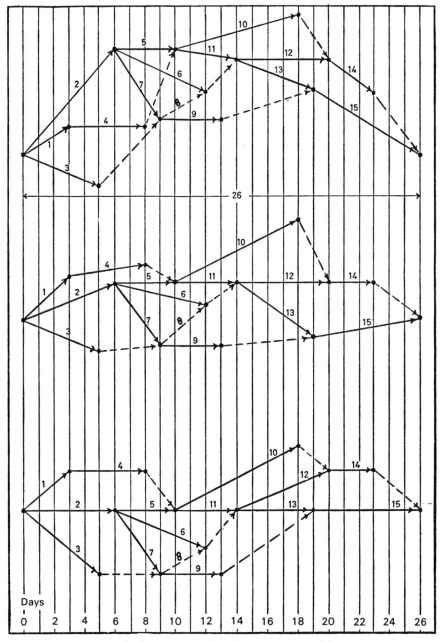

Figure 3.7 Arrow networks drawn to scale. Note that bottom network is inverse of Figure 3.6

25

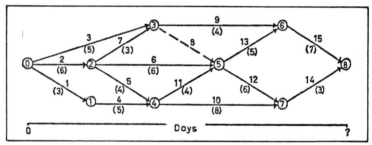

Figure 3.8 Arrow network reduced from Figure 3.6

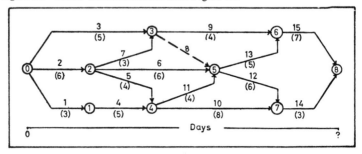

Figure 3.9 Final arrow network

any float attaching to any job. Since the method of determining the critical path is that of cumulating longest path values at the events of the network, these events have to be numbered. This network bears no resemblance at all to the bar chart from which it is derived, and must be subjected to arithmetic procedures before any information can be obtained; the use of computers has been postulated to perform this arithmetic.

The final restriction to be imposed is that by which the arrows of the network shall be drawn as parallel lines. As concurrent activities in a sequence must emanate from their starting event, it seems impossible to show these activities as parallel lines, so the procedure adopted is to contort the arrows of the network so that they conform to a regular pattern. In order to achieve this regularity, most of the activities of a network would be drawn as

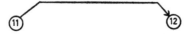

i.e. with the activity now needing three directions of travel. Figure 3.9 shows this type of contortion.

In small networks, such as our example, the shape of the activities does not matter much, but the imposition of 'regularity' as in Figure 3.9 does not add anything of value to a network.

We have mentioned arrow networks quite a lot in these first three chapters, and we have shown that the arrow network adds nothing but complications to project management. However, the practitioner should be reasonably well informed about arrow networks, and the next chapter is devoted to networks in general and arrow networks in particular. It might be interesting, even if it would not be very useful, for the reader to attempt to draw some of the arrow networks for projects described in later chapters and attempt crashing and crash costing from these networks.

Chapter 4

Networks and Arrow Networks

A network is a collection of connected lines, each of which indicates the movement of some quantity between two locations. Generally, entrance to a network is via a *source* (the starting point) and exit from a network is via a *sink* (the finishing point); the lines which form the network are called *links* (or arcs), points at which two or more links meet are called *nodes*. Each particular application of networks has produced its own terminology, and to avoid any possibility of confusion in this chapter the terminology given in this paragraph will be followed closely. Further, the whole system may be called a *graph*, but we shall continue to use the term 'network'.

The mathematics of graph theory has been dealt with admirably in several other books. It is not intended to pursue any mathematics at all in this book since its purpose is to describe a non-mathematical technique. The only requirement is to be able to find the way through a network, from source to sink, and the aim of this chapter is to show that this is really a simple procedure which is well within the capacity of any engineer or industrial supervisor in any other discipline.

Most of the procedures developed to solve complicated networks dealing with flows and pressures through pipes, traffic intensities, etc., require something more than elementary mathematics; but simple networks of the sort discussed in this book, which deal with weights, or volumes, or distances, or times, or any other additive measure, require nothing more than simple addition. This is the only mathematical operation performed in this chapter, and this is the only mathematical operation needed to elucidate a critical path network. Usually the problem is either to determine the shortest path through a network, as when dealing with distances between locations or costs of alternative intermediates in the manufacture of a product, or to determine the longest path through the network as when dealing with performance times. The method employed in each case is to determine the running total value at each node of the network, and a simple network will be worked through to demonstrate each procedure.

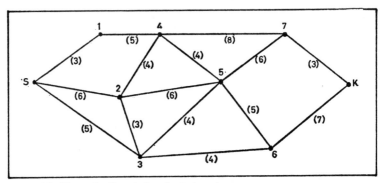

Figure 4.1 Network of roads for shortest route

Take as the first example the network of Figure 4.1 to be a network of roads, with the distances in miles between road junctions (nodes of the network) as marked. The roads are not directed, as each can be travelled in either direction. To find the shortest distance through the network, from S to K, all that has to be done is to tabulate the shortest distance from S to each of the other nodes in turn and select at each node the lowest cumulative total for that node. Thus

$$
\begin{aligned}
\text{from S to 1} &= 3 \text{ miles} \\
\text{1 to 4} &= 5 \text{ miles} \\
\text{so that S to 4, via 1} &= 8 \text{ miles} \\
\text{Alternatively, from S to 2} &= 6 \text{ miles} \\
\text{2 to 4} &= 4 \text{ miles} \\
\text{so that S to 4, via 2} &= 10 \text{ miles} \\
\text{and the shortest distance from S to 4} &= 8 \text{ miles}
\end{aligned}
$$

Continuing through the network in this manner gives the shortest routes as

$$
\begin{aligned}
\text{S to 1} &= 3 \text{ miles} \\
\text{S to 2} &= 6 \text{ miles} \\
\text{S to 3} &= 5 \text{ miles} \\
\text{S to 4} &= 3 + 5 = 8 \text{ miles} \\
\text{S to 5} &= 5 + 4 = 9 \text{ miles} \\
\text{S to 6} &= 5 + 4 = 9 \text{ miles} \\
\text{S to 7} &= 9 + 6 = 15 \text{ miles} \\
\text{S to K} &= 9 + 7 = 16 \text{ miles}
\end{aligned}
$$

so that the shortest route through the network is S → 3 → 6 → K and is 16 miles.

On the other hand, with the network of Figure 4.2 in which it is required to determine the longest route through the network, it is necessary to specify the network discipline, that is a single pass along each link as directed (the arrow on each link indicates the permissible direction of movement along that link) to avoid the possibility of endless cycling. Then for this network, as before, we cumulate distances for each node in turn, but this time we accept the greatest

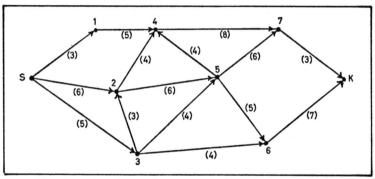

Figure 4.2 Network for longest route

value at each node. The sequence is

S to 1 = 3 miles
S to 2 = 6 miles

but note that node 2 may be reached via node 3 so that

S to 3 = 5 miles

whence

S to 2 via 3 = 5 + 3 = 8 miles
S to 4 via 1 = 3 + 5 = 8 miles, or
S to 4 via 2 = 6 + 4 = 10 miles

But note that node 4 may be reached via node 5 so that

S to 5 via 2   = 6 + 6   = 12 miles, or
S to 5 via 3   = 5 + 4   = 9 miles, or ·
S to 5 via 3 and 2 = 5 + 3 + 6   = 14 miles

which gives us that S to 4 via 3, 2 and 5 = 14 + 4 = 18 miles
               S to 6 via 3   = 5 + 4   = 9 miles, or
               S to 6 via 3, 2 and 5 = 14 + 5 = 19 miles
               S to 7 via 5   = 14 + 6 = 20 miles, or
               S to 7 via 4   = 18 + 8 = 26 miles
               S to K via 6   = 19 + 7 = 26 miles, or
               S to K via 7   = 26 + 3 = 29 miles

Thus the longest single path through the network is that given by

   S → 3 → 2 → 5 → 4 → 7 → K

and is 29 miles long.

When a network is used to represent a project each of the links represents a job. The value given to the link is the duration of the job in minutes, hours, days, weeks, etc., and the overall duration of the project is obtained by calculating the longest path through the network.

As has been demonstrated above, the arithmetic procedure is not difficult, and with the small example network there was very little trouble involved in determining the sequences of nodes. As the network increases in size, possibly with an increase in the number of cross-linkages between nodes, the arithmetic procedure remains the same, but sorting out the sequences of nodes becomes very troublesome and many mistakes are made. The use of computers has been suggested to perform this sorting out and the arithmetic, without the mistakes; alternatively, and preferably, job charts may be used. But let us delve a little into the use of arrow networks for critical path analysis.

Arrow networks for critical path analysis

An arrow network is a diagram constructed to show how all the jobs in a project are related to each other and to the various stages of the project. The principles of construction are the same as those mentioned above, i.e. every network has a starting point (source) and a finishing point (sink) and consists of sequences of links and nodes between these two points. Each link carries an arrow at its head to indicate the direction of movement, hence the name 'arrow network' for a network of directed links. Each job in the project is represented by an arrow in the network extending between two nodes; the tail of the arrow indicates the start of the job, and the head of the arrow indicates the finish of that job. The convention that has been adopted in arrow networks relating to job performance is to use the term *event* for a node of the network and *activity* for a link of the network.

An *event* represents the start or completion of a job and not the actual performance of that job, so that an event has neither work content nor time content; it merely represents a point in time. In the arrow network an event is shown by a small circle or square or ellipse; and since the method of determining the total duration of the project at any event is by cumulating durations at events, it is necessary to identify events. Each event is numbered, and the first event, right at the start of the project (the 'source' of the network) is numbered 0 (zero), and the other events in the network are numbered from 1 upwards. It is not often possible to number the events systematically, but in any case there is no need for this and it should not be attempted.

An *activity* is shown as an arrow drawn between its starting and finishing events, with the arrow head located at and pointing to the finishing event; this arrow indicates the direction of the work flow. Each activity represents a job in the project and so has both work content and time content. However, the arrow (activity) is not drawn to scale. The length given to each arrow is merely a convenient length so that the network fits into the available sheet of paper and also so that each arrow agrees with the general direction of the flow of work. Wherever possible, arrows in the network are drawn so that the flow of work is in the same direction throughout the network, either left-to-right or from the top of the sheet downwards. But in large networks, or networks of activities with many cross-relationships between sequences of jobs, this is not always possible, so that it is important to ensure that the arrows are directed correctly. Each activity is numbered to correspond with the job sheet number of the job it represents, and also is marked with its duration given in the job sheet as standard time. Care is necessary to avoid confusion because the network now carries three series of numbers: the event numbers, the activity numbers, and the activity durations. When discussing the network these should be written in different styles such as ①, ②, ③ etc., for event numbers; 1, 2, 3 etc., for activity numbers; and (5), (15), (25) etc., for activity durations.

It is very unlikely that all the activities in a project will lie in one sequence, so that the jobs can be completed one after the other without hindrance; such an occurrence is the simplest of projects. Usually we find that a project has several sequences of activities, and there may be some relationships between activities in different sequences. These relationships are time-relationships and must be shown in the arrow network, because a job in one sequence may depend for its start on the completion of jobs in other sequences. 'Dummy' activities are needed to show these relationships; the dummies have neither work content nor time content, and are drawn in the arrow network solely to identify the relationships and permit them to be included in the summation of durations through the network. Each dummy must be directed to show which way the relationship connects the activities. For example, part of a project might consist of the following jobs

    Job 8 : place laboratory benches,
    Job 9 : fit cupboards to benches,
    Job 10 : fit gas and water taps to benches,
    Job 11 : install gas and water pipes in laboratory,
    Job 12 : connect pipes to bench taps,
    Job 13 : connect pipes to mains,

which, because of the work schedule of other parts of the project, must be carried out in this order. That is, Job 13 must follow Job 12, maybe at a much later date, although under other circumstances Job 13 might precede Job 12 (if the mains were installed before the laboratory distribution pipes).

It is obvious that when Job 8 has been finished both Jobs 9 and 10 may be carried out concurrently. Job 11 is independent of all of Jobs 8, 9 and 10 and may be carried out concurrently with these jobs. Job 12 may be started only when *both* Jobs 10 and 11 are finished. Job 13 follows Job 11 and is independent of Jobs 10 and 12. Then a network such as Figure 4.3 is obviously wrong because it shows that the starts of both of Jobs 12 and 13 are dependent on the completion of both of Jobs 10 and 11, whereas it is clear that Job 13 is nothing to do with Job 10 and can have no dependence on it. Jobs 11 and 13

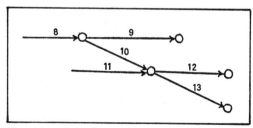

Figure 4.3 Flow of work

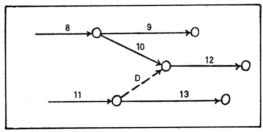

Figure 4.4 Flow of work showing position of a 'dummy'

must be drawn in one sequence and a dummy inserted to show that the start of Job 12 is dependent on the completion of Job 11. Figure 4.4 shows the correct arrangement of activities for this part of the project.

Determining the correct position and direction of dummies is reputed to be one of the more difficult factors in the construction of an arrow network. Each dummy must be dealt with very carefully; its position and direction must be correct, as an error made with a dummy can vitiate the whole arrow network and the calculations made from it.

Arrow network from the project of Figure 2.1

An arrow network is constructed from the data provided in the job sheet and, except for very simple projects involving only a few jobs, an arrow network is not easy to construct. Figure 4.5 shows an arrow network constructed from the data of the job sheet given in Figure 2.1, and a brief description of the method of construction may indicate some of the intricacies involved.

The job sheet shows that three jobs, Jobs 1, 2 and 3, may be started at the beginning of the project. The starting event of the project, ⓪, is written at the left-hand side of the page and the three activities 1, 2 and 3 are drawn radiating from this event. These are numbered 1, 2 and 3, respectively, and the appropriate job durations are marked on them. Their finishing events are numbered ①, ② and ③, respectively. On completion of Job 1, Job 4 may be started, so activity 4 is drawn from event ① to terminate at event ④, is numbered and its duration marked.

On completion of Job 2, all of Jobs 5, 6 and 7 may be started, so that

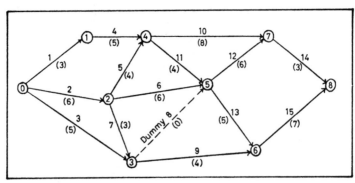

Figure 4.5 Arrow network for the project of Figure 2.1

32

activities 5, 6 and 7 may be drawn from event ②. But looking down the job sheet we see that there is no job which follows from Job 4 only or from Job 5 only; but there are two jobs, Jobs 10 and 11, which may be started on completion of *both* Jobs 4 and 5. Therefore, activity 5 is drawn from event ② to event ④, is numbered and its duration marked; event ④ now denotes the completion of *both* of Jobs 4 and 5. We see also that there are two jobs, Job 12 and 13, which may be started on completion of a collection of jobs including both Job 6 and Job 7, whereas another job, Job 9, may be started on completion of *both* Jobs 3 and 7. Accordingly, activity 6 is drawn from event ② ending in event ⑤, is numbered and its duration marked. Activity 7 is drawn from event ② to event ③, is numbered and its duration marked; event ③ now denotes the completion of *both* of Jobs 3 and 7.

Now activity 9 can be drawn from event ③ ending in event ⑥, and its number and duration marked. We see that *both* Jobs 10 and 11 may be started when *both* Jobs 4 and 5 are complete, so that activities 10 and 11 will be drawn from event ④. However, the job sheet shows that the completion of Job 10 has to be related to the completion of Job 12 for Job 14 to start, while the completion of Job 11 must be related to the completion of Jobs 3, 6 and 7, all of which must be completed before Jobs 12 and 13 can be started. Activity 10, therefore, is drawn from event ④, ending in event ⑦, is numbered and its duration marked. Activity ⑪ is drawn from event ④ to event ⑤, is numbered and its duration marked; event ⑤ now denotes the completion of *both* of Jobs 6 and 11.

Both Jobs 12 and 13 can be started when Jobs 3, 6, 7 and 11 are *all* complete. We have Jobs 3 and 7 related to Job 9 in one sequence, while Job 9 is not related to Jobs 6 and 11. Thus a dummy activity is necessary to bring together the finishing points of Jobs 3, 6, 7 and 11, while the finishing points of Jobs 3 and 7 may be kept separate from the finishing points of Jobs 6 and 11. The dummy is drawn from event ③ to event ⑤, and event ⑤ now denotes the completion of all four Jobs 3, 6, 7 and 11. Now both activities 12 and 13 can be drawn from event ⑤; but we note that *both* Jobs 10 and 12 have to be completed before Job 14 can be started, so that activity 12 is drawn from event ⑤ to event ⑦, is numbered and its duration marked. Event ⑦ now denotes the completion of *both* of Jobs 10 and 12. Also, we see that *both* Jobs 9 and 13 have to be completed before Job 15 can be started, so that activity 13 is drawn from event ⑤ to event ⑥, is numbered and its duration marked; event ⑥ now denotes the completion of *both* of Jobs 9 and 13.

Finally, we see that Jobs 14 and 15 are the last jobs in the project so their corresponding activities must end at a common event. Activity 14 is drawn from event ⑦ ending in event ⑧, is numbered and its duration marked; activity 15 is drawn from event ⑥ to event ⑧, is numbered and its duration marked. Event ⑧ denotes the completion of both of Jobs 14 and 15; it is the finishing event of the arrow network and represents the end of the project.

In an arrow network the arrows are not drawn to scale, so that even when constructed successfully the arrow network is of little use since it contains no information that has not been provided already in the job sheet. In order to derive additional information — the information that is really required — it is necessary to translate the job sheet and arrow network into a procedure which will determine the critical path and provide any other information that may be wanted. Two such procedures are described.

Critical path matrix

The first procedure entails construction of a critical path matrix, also known as a job matrix. Compilation of the job matrix, also, is a somewhat difficult procedure, and compilation of a matrix for even a simple project requires considerable thought and skill. Possibly familiarity breeds contempt and the procedure becomes less fraught with difficulties with frequent use. Once compiled, the matrix enables critical jobs to be identified, but the matrix must be recalculated for every modification proposed and a computer may be needed to deal with the arithmetic involved. The job matrix deals with the events of the arrow network and not its activities; the durations of activities are entered in the matrix as durations between starting events and their corresponding finishing events. For those who may be interested, compilation of a job matrix is as follows.

Let the start and finish of each activity be designated by i and j respectively, so that every activity is determined by reference to these letters and the terminal events of the activity. Remember that the events are numbered and the activities are numbered, and care must be taken at this stage to avoid confusion between the two series of numbers. Taking an example from the project job sheet of Figure 2.1 and comparing with the arrow network of Figure 4.5, we see that Job 4 of the job sheet is represented by activity 4 of the arrow network; this activity extends from the i (starting) event ① to the j (finishing) event ④ and thus may be written as ($i1$, $j4$). Similar representations for the other activities enable the job matrix to be compiled (see Figure 4.6).

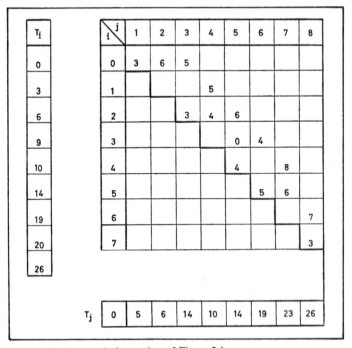

T_i		i \ j	1	2	3	4	5	6	7	8
0		0	3	6	5					
3		1				5				
6		2			3	4	6			
9		3					0	4		
10		4						4	8	
14		5						5	6	
19		6								7
20		7								3
26										
	T_j	0	5	6	14	10	14	19	23	26

Figure 4.6 Job matrix for project of Figure 2.1

The matrix is constructed with i rows and j columns, and each entry in the matrix represents an activity. The duration of each activity is entered in the appropriate cell of the matrix; for example, activity 10 has a duration of 8 days and extends from event ④ to event ⑦, so the value 8 is entered in cell ($i4, j7$). All dummies must be shown in the matrix, and as these have no duration each is given a value of zero. All entries in the job matrix should be above the diagonal. An entry below this diagonal means that the work flow is in a negative direction and the plan needs correction. To use the matrix for determination of the critical path, a summation column T_i, which gives the total duration to the start of any given activity, and a summation row T_j, which gives the total duration to the finish of any given activity, are added, and the summation of each is started from zero. The summation values T_i must be made in close conjunction with the arrow network; the T_i values are, of course, the maximum path values at each event. By definition, at event ⓪ no work has been done, so that T_i at $i0$ is zero. Values of T_i for other events are obtained by straightforward cumulation. The summation values T_j are obtained by a reverse count (maybe 'reverse subtraction' is a more correct term). The last entry of 26 is made for T_j ($j = 8$), and activity durations are subtracted in sequence as indicated by the arrow network. For example, T_j at event ④ is

$$T_j (j = 4) = 26 - 7 - 5 - 4 = 10$$

The matrix is completed as shown in Figure 4.6, and the critical path is determined by considering the differences $T_i - T_j$ for each event. Where the difference is zero, this means that the latest finishing time for an activity which ends at that event is the same as the earliest starting time for the next activity which begins at that event, and thus the event lies on the critical path.

Once the critical path has been determined it is possible to determine the relationship of each job to the critical path. In our example project the critical path connects events ⓪, ②, ④, ⑤, ⑥ and ⑧ and consists of activities 2, 5, 11, 13 and 15. All other jobs, or sequences of jobs, have some float. For example, Job 10, which requires 8 days for completion, is represented by activity 10 extending from event ④ to event ⑦ and has a total time availability of $T_i7 - T_i4 = 10$ days, so that Job 10 has a free float of $10 - 8 = 2$ days.

Forward and backward count

A simpler method of identifying the critical activities of an arrow network, by means of their related events, is to make a longest path count from each end of the network and total the counts at each event. This is a much simpler procedure to use than the job matrix. The difference between the job matrix and the longest path count method is that the job matrix method determines those events at which the latest finishing time of an activity must be the same as the earliest starting time of the next activity, while the longest path count method determines those events which lie on the longest continuous path — the critical path — through the network. Figure 4.7 shows the counts for our example network.

Event :	0	1	2	3	4	5	6	7	8
Forward :	0	3	6	9	10	14	19	20	26
Backward :	26	21	20	12	16	12	7	3	0
	26	24	26	21	26	26	26	23	26
	x		x		x	x	x		x

Figure 4.7 Longest path count

All the arithmetic can be eliminated by drawing the arrow network to scale, and three possible structures for this network were shown in Figure 3.7. It is more difficult to draw an arrow network to scale than to draw just an arrow network, and, as may be seen in Figure 3.7, a network may be made to appear reasonably simple or deliriously complicated. In scale networks the critical path appears as a continuous line through the network, but resolution of non-critical paths is still very difficult and the simple representation of projects in the form of job charts is recommended.

Chapter 5

Crashing the Project

In every organisation some of the repair or replacement work is rushed through under emergency conditions, when all available effort is brought to bear to complete the work in the shortest possible time. Usually the duration of the emergency work is very much less than the duration would have been if the work had been performed under standard conditions. The reduction in time is accompanied by an increase in cost, but the increased cost is not often considered; first priority is given to getting the plant back in operation. This idea of 'crashing' a job can be introduced into the planning stage of a project; but instead of crashing all jobs, due consideration must be given to the fact that completion of the job in less than the standard time is feasible and will be included in the plan if, and only if, the higher cost for the job can be tolerated. The higher cost can be generated by the use of additional labour — more men on the job or the same men working overtime, or by using contractors; premium payments for quicker delivery of materials; hiring extra equipment or tools. It is necessary to determine

1   Which jobs in a project should be crashed.
2   What benefit is likely to accrue from crashing.

From the answers it will be possible to obtain the maximum saving in time for the minimum additional expenditure of money, if it can be demonstrated that it is worthwhile spending this additional money.

Crash times and their associated crash costs for the example project are included in the job sheet of Figure 2.1, and these are the only feasible crash times for these jobs. It may be noted that if all the jobs were crashed indiscriminately the total project cost would be £10,900. A decision must be made as to which jobs should be crashed, but neither the job sheet nor the arrow network provide any assistance in reaching this decision. However, the job chart of Figure 3.1 shows which jobs should be crashed to effect some saving of time, and also shows which should not be crashed. It is obvious from Figure 3.1 that crashing some jobs, such as Jobs 3 and 9, will not result in any reduction in the overall project duration. Such jobs as these would be crashed only if the manpower or some other resource were in short supply, when it becomes essential to complete the jobs as quickly as possible and transfer the resource to a more important job in this or another project. Otherwise, crashing such jobs involves a needless expenditure of money.

The job chart of Figure 3.1 shows the critical path to consist of Jobs 2, 5, 11, 13 and 15, and the job sheet of Figure 2.1 shows the time savings by crashing these jobs to be 2, 2, 2, 1 and 2 days respectively. Then if no other job becomes critical and interferes with the proposal, it should be possible to reduce the project duration by 9 days, that is to 17 days. This can be done and Figure 5.1 shows the job charts for minimum project time. Chart 1 shows the plan if all jobs were to be crashed regardless of necessity, while Chart 2 shows the plan with only those jobs crashed which lead to a reduction in the project duration. From Chart 1 we see that there is sufficient float attaching to Jobs 3, 7, 9 and 14 to obviate any thought of crashing these jobs; as far as time alone is concerned, crashing these jobs is of no help in reducing the project duration. The sequence of Jobs 1 and 4 has a one-day float, but the job sheet shows that this 1 day can be utilised only on Job 1 so that Job 4 will need to be crashed. Also, we see that there is a 3-day float after Job 14, which was crashed by only 1 day, so that if 1 day of this 3-day float be returned to Job 14, the other 2 days' float can be used to return Jobs 10 and 12 to their standard times. These two jobs are concurrent and the 2 days' float from Job 14 becomes available to *each* of Jobs 10 and 12. Thus the minimum project duration can be achieved, for the present disregarding any other considerations, with Jobs 1, 3, 7, 9, 10, 12 and 14 remaining at their standard times.

Several methods can be used to derive the job chart for minimum project duration with selective crashing. As in many other procedures a systematic approach is best, and this is adopted for our example project. It is necessary to consider jobs on the critical path and determine what effect on the project duration can be made by altering the duration of the critical jobs. Then from the job chart of Figure 3.1 and the crash data in the job sheet of Figure 2.1, we see that

1   Job 2 can be reduced from 6 days to 4 days without affecting any other job. If this be done, the time line at the end of Job 2 is raised by 2 days, and the start and finish of every job dependent on this time line are advanced by 2 days. For this project this means every job except Jobs 1, 3 and 4. The time line at the end of Job 5 now meets the end-line of Job 4, wiping out the float that existed here in the original plan; and the time line at the end of Job 7 reduces the Job 3 float from 4 to 2 days. The critical path duration is reduced from 26 to 24 days.

2   Job 5 can be reduced from 4 days to 2 days, but this means that 2 days must be saved between Jobs 1 and 4. Job 1 can offer a saving of only one day, whereas Job 4 can offer the required saving of two days; so Job 4 is crashed from 5 to 3 days, and Job 1 is allowed to remain at its standard time of 3 days. The time line at the end of Jobs 4 and 5 is raised by 2 days, and the start and finish of every job dependent on this time line are advanced by 2 days. This moves the time line at the end of Job 11 so that it meets the end of Job 6 and wipes out the float that existed here in the original plan; note that if Job 6 had less than 2 days' float it would have been necessary to crash Job 6 also. The

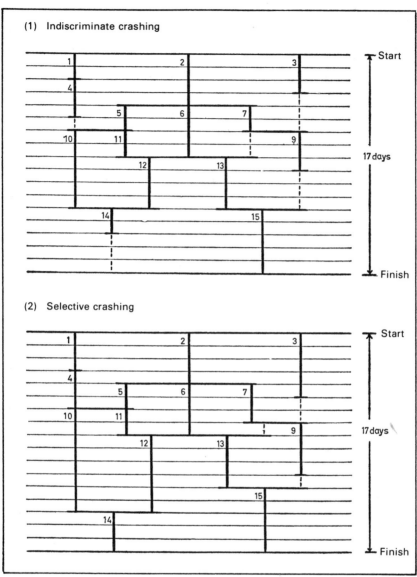

Figure 5.1 Job charts for minimum overall project duration

gap between the time line at the end of Job 7 and the time line at the end of Job 11 is reduced from 5 to 3 days. The float following Job 9 is reduced from 6 to 4 days, and the critical path duration is reduced from 24 to 22 days.

3   Job 11 can be reduced from 4 to 2 days, but to effect any saving in time Job 6 must also be reduced by 2 days. The job sheet shows that this is feasible, so both Jobs 6 and 11 are reduced by 2 days. The time line at the end of Jobs 6 and 11 is raised by 2 days, and the start and finish of every job dependent on this time line are advanced by 2 days. The time line at the end of Job 12 now

meets the end-line of Job 10 and wipes out the float that existed here in the original plan; again note that if Job 10 had less than 2 days' float it would have been necessary to crash Job 10 also. The gap between the time line at the end of Job 7, and the time line at the end of Job 11 is reduced from 3 days to 1 day. The float following Job 9 is reduced from 4 days to 2 days, and the critical path duration is reduced from 22 to 20 days.

4   Job 13 can be reduced from 5 to 4 days without affecting any other job. The time line at the end of Job 13 is raised by 1 day, and the start and finish of Job 15 are advanced by 1 day. The float following Job 9 is reduced from 2 days to 1 day; and the project finishing time line at the end of Job 15 is advanced by 1 day, so that the float following Job 14 is reduced from 3 to 2 days. The critical path duration is reduced from 20 to 19 days.

5   Finally, Job 15 can be reduced from 7 to 5 days without affecting any other job. The finish of the project is advanced by 2 days, and this wipes out the float following Job 14. Had this float been less than 2 days it would have been necessary to crash Job 14 as well. The critical path duration is reduced from 19 to 17 days, and this is the minimum project duration as there are no more critical jobs whose durations can be shortened to effect a further reduction in the overall project duration.

Thus, as suggested when we considered Chart 1 of Figure 5.1, the minimum project duration has been achieved by crashing only Jobs 2, 4, 5, 6, 11, 13 and 15, leaving Jobs 1, 3, 7, 9, 10, 12 and 14 at their standard durations. The final job chart, Chart 2 in Figure 5.1, is the chart for minimum project duration, and this shows that Jobs 1, 2, 4, 5, 6, 10, 11, 12, 13, 14 and 15 are now critical.
The minimum project duration chart has been achieved by considering the critical jobs, and the possibilities attached to crashing them, in order from the start of the project, i.e. deal with the jobs in the order in which they form the critical path: Jobs 2, 5, 11, 13 and 15. Consideration of these jobs in reverse order or in any random order would have achieved the same final result; but systematic consideration of the critical jobs as they arise from start to finish of the project is thought to be the easiest procedure, and we have adopted this procedure in all our preparations of minimum duration project plans. Where job costs are to be considered, a different order of priority may be forced on the planner, and this is dealt with in the next chapter. In Figure 2.2 we showed how the critical path, project duration, and job floats could be derived from the job sheet, and sometimes this procedure could be useful; but attempts to derive the minimum duration project from the job sheet alone are thought to be most unrewarding because working from the job chart is so easy. However, examination of the job sheet will sometimes give an indication of which jobs should not be crashed, and we can look at the job sheet for a different project and consider some of the moves that can be made. In Figure 5.2

1   We see that Job 2 and Jobs 1, 3 and 4 form a parallel sequence. Job 2 duration cannot be reduced, and so there is need only to reduce the total duration of Jobs 1, 3 and 4 to 4 days. This is best accomplished by reducing Job 1 duration from 3 days to 1 day, since only 1 day can be saved on Job 3 and none on Job 4.

2   Next we have another parallel sequence controlled by Jobs 5 and 7 on

Job Number	Job Description	Job Sequence	Standard Duration	Standard Cost	Crash Duration	Crash Cost
1	Lead time for management approval	0: 1	3 days	£ —	1 day	£ —
2	Line availability	0: 2	4	—	4	—
3	Measure and sketch	1: 3	2	60	1	80
4	Develop materials list	3: 4	1	20	1	20
5	Procure pipe and flanges	4: 5	5	170	3	220
6	Procure valves	4: 6	9	60	5	120
7	Prefabricate sections in workshop	5: 7	5	240	2	400
8	Deactivate line	2,4: 8	1	20	1	20
9	Erect scaffold	2,4: 9	2	60	1	100
10	Remove old pipe and remove to scrapyard	8,9:10	4	80	2	200
11	Place new pipe in position	7,10:11	6	300	3	600
12	Weld pipes	11:12	2	20	1	60
13	Place valves in position	6,8,9:13	2	20	1	50
14	Fit up valves and pipes	12,13:14	2	20	1	50
15	Pressure test on system	14:15	2	10	1	20
16	Insulate	12,13:16	4	80	2	140
17	Remove scaffold	14,16:17	2	20	1	50
18	Clean up	15,17:18	2	20	1	50
		Project	?	£1200	?	?

Figure 5.2 Job sheet for the project 'Renewing a pipeline'

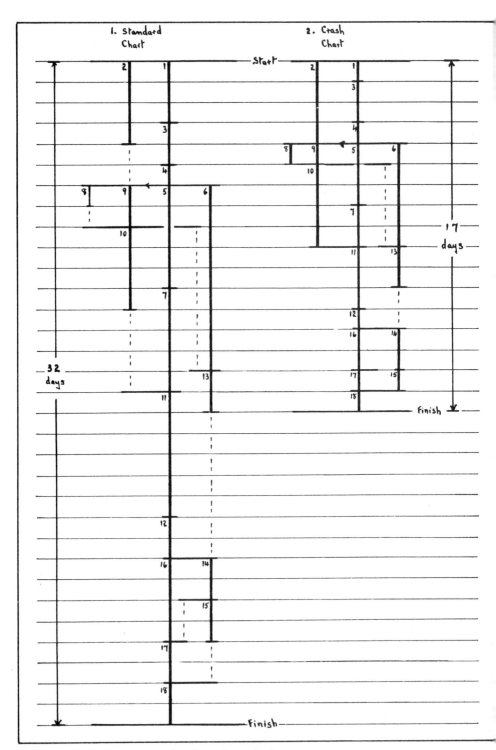

Figure 5.3 Job charts for the project 'Renewing a pipeline'

42

one side and Jobs 8, 9 and 10 in the other. If Job 5 is crashed to 3 days and Job 7 to 2 days, then we need to reduce the chain of Jobs 8, 9 and 10 (which are not all in sequence) to 5 days, and this seems to be best achieved by crashing Job 9 from 2 days to 1 day. Job 8 cannot be crashed, and since Jobs 8 and 9 are concurrent, starting from the finish of Job 2, there is no need to crash Job 8; while if we crashed Job 10 we would save 2 days and thereby create 1 day's float.

3   Now it is a little difficult to see what comes next, but if we look at the jobs up to the start of Job 14 we see that we have two parallel sequences: if we were to crash Job 11 to 3 days and Job 12 to 1 day, one sequence would consist of Jobs 5, 7, 11 and 12 and would extend over 9 days, while the other sequence consists of Jobs 6 and 13. The standard times for these jobs total 11 days; Job 13 can be reduced only by 1 day, which is not enough, while Job 6 can be reduced by 4 days, which really is too much. But Job 6 must be reduced to 5 days and Job 13 left at standard so that there are 2 days' float after this sequence of jobs.

4   Finally, we see that Job 16 can be crashed to 2 days, so that each of Jobs 14 and 15 will have to be crashed to 1 day, and Jobs 17 and 18 can also be crashed to 1 day each.

Hopefully, the above description fits the facts; the charts in Figure 5.3 show what must really happen. But there can be no doubt that, for any project with a large number of cross-relationships, attempting to derive a crash schedule from the job sheet would be very time-consuming and could not possibly be worthwhile.

Chapter 6

Project Costing

When the first feasible plan for a project has been prepared, the next item for consideration is the possibility of reducing the overall project duration, and this reduction is deduced as described in Chapter 5. However, unless there is some over-riding requirement that a project be completed in the shortest possible time, in which case the plan for minimum duration becomes the final work plan subject to resource scheduling, the jobs must be considered in a systematic manner; and the time saving on any job or collection of associated jobs must be related to the premium expenditure demanded. The object of this part of the planning procedure is to determine how the maximum time may be saved for minimum premium expenditure, and the order of crashing jobs that may be recommended finally is not necessarily the order we considered previously. Firstly, each job which can exert an individual effect on the project duration should be considered, after which it becomes necessary to consider any combination of jobs which exerts a single effect. A tabulation of jobs in ascending order of premium expenditure per day saved is prepared so that it is easy to determine the maximum time which may be saved for any agreed level of premium expenditure.

It is obvious from the job sheet of Figure 2.1 and the job chart of Figure 3.1 that:

1   Job 15 can be reduced from 7 to 5 days without affecting any other job. The premium expenditure that has to be incurred to permit this saving of time is £350, i.e. a premium rate of £175 per day saved.

2   Job 13 can be reduced from 5 to 4 days without affecting any other job. The premium expenditure that has to be incurred to permit this saving of time is £100, i.e. a premium rate of £100 per day saved.

3   Now the possible reductions to be considered are: Job 2 from 6 to 4 days; Job 5 from 4 to 2 days; and Job 11 from 4 to 2 days. There are three possible combinations to effect these reductions, as it will be necessary to

crash Jobs 4 and 6 with one or more of these jobs. The possible combinations are:

a   Job 2; Jobs 4 and 5; Jobs 6 and 11.
b   Jobs 2 and 4; Job 5; Jobs 6 and 11.
c   Job 2; Job 11; Jobs 4, 5 and 6.

The premium expenditures for these combinations are:

a   Job 2 can be crashed by 2 days at a premium expenditure of £250, i.e. a premium rate of £125 per day. Jobs 4 and 5 can be crashed by 2 days at a premium expenditure of £600, i.e. a premium rate of £300 per day. Jobs 6 and 11 can be crashed by 2 days at a premium expenditure of £400, i.e. a premium rate of £200 per day.

b   Jobs 2 and 4 can be crashed by 2 days at a premium expenditure of £500, i.e. a premium rate of £250 per day. Job 5 can be crashed by 2 days at a premium expenditure of £350, i.e. a premium rate of £175 per day. Jobs 6 and 11 can be crashed by 2 days at a premium expenditure of £400, i.e. a premium rate of £200 per day.

c   Job 2 can be crashed by 2 days at a premium expenditure of £250, i.e. a premium rate of £125 per day. Job 11 can be crashed by 2 days at a premium expenditure of £200, i.e. a premium rate of £100 per day. Jobs 4, 5 and 6 can be crashed by 2 days at a premium expenditure of £800, i.e. a premium rate of £400 per day.

If all five jobs are to be crashed for minimum project duration it does not matter how they are combined; the total premium expenditure on the five jobs is £1,250. But if there is any likelihood that the premium rate might become an important factor in determining the maximum saving of time for some maximum permissible premium expenditure rate, then combination (c) is accepted and this part of the programme is arranged as:

Crash Job 11 by 2 days at a premium rate of £100 per day.
Crash Job 2 by 2 days at a premium rate of £125 per day.
Crash Jobs 4, 5 and 6 by 2 days at a premium rate of £400 per day.

The final arrangement for the project may be shown as in Figure 6.1 with premium expenditures per day saved, now called the 'cheapness ratings', given in ascending order.

These cheapness ratings, or premium expenditures per day saved, must be compared with the loss of profit due to non-availability of the equipment

Job	Days saved	Premium Expenditure	Premium Expenditure per day saved
13	1	£100	£100
11	2	200	100
2	2	250	125
15	2	350	175
4, 5 & 6	2	800	400
Total	9	£1,700	£189

Figure 6.1   Evaluation of job crashing

concerned in the project before a decision can be made regarding whether or not any job should be crashed. In this example, if the expected profit were less than £100 per day then none of the jobs would be crashed as the premium expenditure cannot be justified; if the profit were £150 per day then Jobs 2, 11 and 13 would be crashed; and if the profit were £400 per day or more then all seven of the jobs would be crashed. If all these jobs were crashed, the total project cost would be £9,600 for the 17-day duration project, £1,300 less than would be the total project cost if all the jobs in the project were to be crashed indiscriminately. And note that the project average cheapness rating shown in the 'Project total' line of Figure 6.1 should not be used as an argument for crashing. For instance, if the value of the project were £200 per day, it would be immoral and most unprofessional to crash all seven jobs on the grounds that 'on average' there is a saving. The whole purpose of the planning exercise is to maximise profit, and no crashing at a cheapness rating higher than the project value shall be permitted unless enforced by resource allocation between projects.

Where overheads or other indirect expenses may be charged against the project, it is possible to determine that project duration which can be achieved for minimum total project cost. Whereas the direct costs of the project are costs of the work involved in each job, usually the overheads are chargeable at a fixed rate per day regardless of how much work is done in any particular day. For instance, in our example project, if overheads at £150 per day were to be charged against the project, the 21-day project would be the project that shows the minimum total project cost. But even if a minimum total project cost is calculated, the cost of crashing any other jobs must be compared with the estimated loss of profit. The loss of profit is the response that has to be minimised; overheads always seem to be paid however they are allocated.

A project-cost tabulation can be prepared, as shown in Figure 6.2, showing how direct costs and total costs vary with the project duration. Note that as the overheads per day increase, the Total Project Cost minimum moves downwards through the table. At £200 per day for overheads, the 19-day project has the minimum total project cost, while the 17-day project does not become the minimum until overheads reach £400 per day.

Cost-duration curves can be drawn from the tabulated data, and these are shown in Figure 6.3. If curves are drawn for a project, the cost-duration curve constructed from the direct costs is the more important as this deals with the controllable costs of the project. It must be remembered that, in general, only the plotted points have real existence; the curve has no meaning between these

Project duration	Direct project cost	Overheads of £150 per day	Total project cost
26 days	£7,900	£3,900	£11,800
25	8,000	3,750	11,750
23	8,200	3,450	11,650
21	8,450	3,150	11,600
19	8,800	2,850	11,650
17	9,600	2,550	12,150

Figure 6.2   Project-cost tabulation

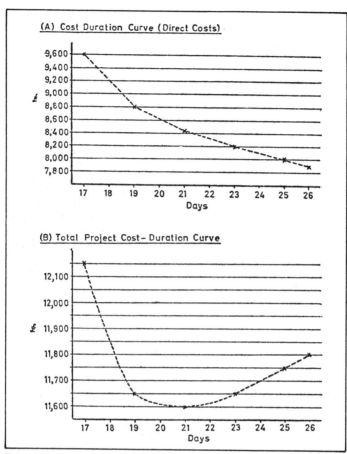

Figure 6.3 Cost-duration for the project of Figure 2.1

points and is drawn between the plotted points only to demonstrate the upward trend of the project cost as the project duration decreases. In the example project each of Jobs 11 and 13 can be crashed at a premium expenditure of £100 per day, so that there is no absolute need to consider Job 13 as the first job to be crashed; Job 11 may be put first in the tabulation when a project duration of 24 days becomes feasible. The premium expenditure per day necessary to move from one point on the cost-duration curve to the next feasible point, which we have called the cheapness rating (it could be called the premium rate) may be called the 'cost slope'; and in our example there is a constant slope of 100 between the 26-day project and the 23-day project, with feasible intermediate points at 25 and 24 days. Similarly, the total project cost-duration curve exists only at the plotted points. It may be possible to draw a curve through the plotted points which would have a minimum at some duration other than a plotted point, and this is not permissible. In the curve shown in Figure 6.3.2, it seems that a curve could be drawn to show a minimum at $20\frac{1}{2}$ days. This must not be done; there is no $20\frac{1}{2}$-day project. To avoid the possibility of showing a non-existent minimum, it is better to join successive plotted points by straight lines and not attempt to draw a curve.

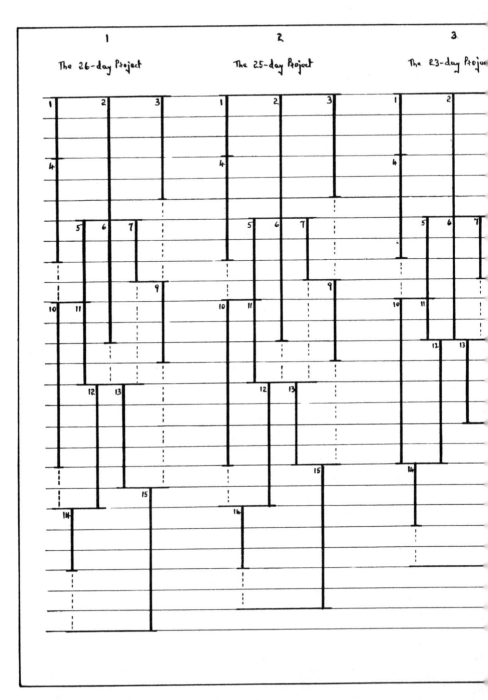

Figure 6.4 Project crashed in steps as required by Figure 6.1

48

4

The 21-day Project

5

The 19-day Project

6

The 17-day Project

49

Figure 6.4 shows the modifications made in the job chart if the project duration is reduced in steps according to the order listed in Figure 6.1. The effects of each reduction on the other jobs and on the floats is clearly seen.

Chapter 7

Manpower and Resource Allocation

In the preceding chapters, taking a small project as an example, we have dealt with the planning of the work involved in the project and analysis of a feasible plan to provide a better one, where 'better' is defined as less time-consuming but still economically acceptable. Firstly, a plan was prepared of all the work that had to be completed before the project could be accepted as 'finished'. This plan required preparation of the job sheet, which listed the relevant details of every job in the project, and of the job chart, which showed the relationships between these jobs; this job chart was the first feasible plan for the project. The job sheet and job chart provided all the available information about the expected duration of jobs, the necessary sequencing of these jobs, and the costs of getting the jobs done. Secondly, an analysis was begun of this first feasible plan. The critical jobs were identified, and the length of the critical path through the project was determined; this showed the expected overall project duration and the jobs which controlled this duration. Then crashing the critical jobs was considered and the likely effects on the duration and the costs of the project that would be achieved. It was found that the project duration could be reduced progressively by crashing the critical jobs in a systematic manner from start to finish of the project. It was also observed that crashing some of the critical jobs had to be accompanied by crashing some jobs which up to that time had been non-critical; otherwise these jobs would have become critical and would have prevented a reduction in the project duration. The final job chart for minimum project duration, as shown in Figure 5.1.2, can be proposed as a reasonable start for the project scheduling, and if this is accepted starting and finishing dates can be inserted for the project and for each job in the project. Now we can continue with the job chart of Figure 5.1.2 to schedule manpower, or any other resource, to meet the requirements of the project. To permit an extension of the job chart for use in manpower and resource allocation and control, the manpower and resource requirement of every job is listed alongside the relevant job line on

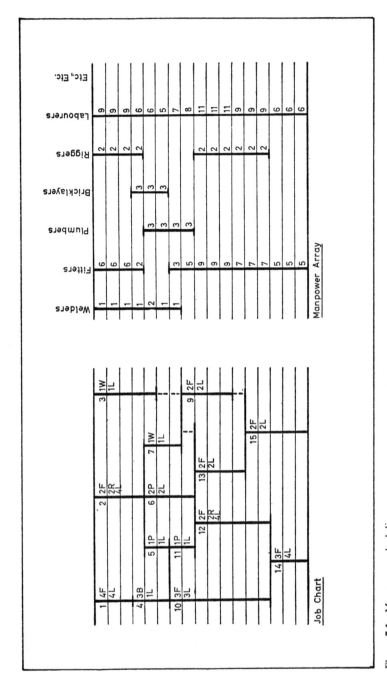

Figure 7.1 Manpower scheduling

52

the job chart; for most of the projects manpower is the only resource that needs scheduling, so 'manpower' and 'other resources' are separated. A similar array of vertical lines is constructed, giving each line its appropriate heading: 'Welders', 'Fitters', etc. On each vertical line of the manpower and resource array, the total requirement for each interval of time is listed. The total requirement in any interval of time is, of course, the total of the individual job requirements in that interval of time, and is obtained by summing across the job chart. Some job lines on the job chart may correspond to more than one manpower or resource line; for instance, it is easy to visualise a job on which fitters, riggers and labourers are engaged at the same time, and for such a job the one job line on the job chart would require three entries in the manpower array. Also, several job lines in the job chart may be amalgamated into one line in the manpower and resource array; and in this case it might be convenient to divide the one manpower or resource line into sections to correspond with the requirements of the individual job lines shown in the job chart, though this has never been needed to date. When the array is complete, the requirement of each resource can be compared with the known availability of that resource, and any necessary amendments may be made to the job chart so that requirement and availability agree in every case.

Although an array can be enlarged to show the requirements of all resources to be used in a project, there is no need to do this and include any resource which does not impose a restriction on this or some other project. Unnecessary complications in the array construction should be avoided. The manpower and other resources array is likely to be of greater use when a number of projects have to be fitted together than when a single project is being considered on its own. In this case it would be convenient to include in all the project arrays details of any resource which might be restrictive in just one project, so that amendments may be made most easily and with least disturbance to the overall scheduling. As soon as the manpower and other resource requirements are known, scheduling can begin in earnest.

For the project taken as the example it was necessary to consider only the availability of the various craftsmen; there was no restriction from any other resource. Figure 7.1 shows the job chart for minimum project duration with the manpower requirements for each job listed against the appropriate job line; each type of craftsman is indicated by the initial letter of the name used in the manpower array. The manpower array shows the total requirement of each craft for every day of the project under six headings, and these shall be considered separately.

Every job in the project requires effort from labourers, and the 'Labourers' line in the manpower array shows the variation in requirement through the project; the extremes are 5 men on the sixth day and 11 men on each of the ninth, tenth and eleventh days. Riggers are needed on two jobs, and the two sections of the 'Riggers' line correspond to Job 2 (upper section) and Job 12 (lower section). Bricklayers are required on only one job, so that the short 'Bricklayers' line corresponds to Job 4. Plumbers are required on three jobs which have been planned (tentatively?) so that two jobs are carried out concurrently, firstly Jobs 5 and 6, and then Jobs 11 and 6; the 'Plumbers' line therefore corresponds to Jobs 5, 6 and 11. Fitters are required on eight of the jobs, and the two sections of the 'Fitters' line correspond to Jobs 1 and 2 (upper section) and Jobs 9, 10, 12, 13, 14 and 15 (lower section). All the craftsmen required for these jobs can be made available at the times indicated in the job chart, and there is no need to do any further scheduling on their behalf. If in any instance the required number of craftsmen could not be made available, the job chart and manpower array would have to be manipulated to provide the necessary balance, and this would extend the project duration. For example, if only two plumbers could be made available at any time, then obviously Jobs 5 and 6 and Jobs 11 and 6 could not be carried out together and a re-arrangement would be necessary. These three jobs would have to be tackled in the order 5, 6 and 11, or 6, 5 and 11, according to which days the two plumbers can be made available; and whichever order is used means an extension of four days to the project duration.

Finally, the job chart shows that the services of a welder are needed for each of Jobs 3 and 7, and as drawn in the job chart each of these jobs requires the services of a welder on the fifth day of the project. This is not feasible as only one welder can be allocated to the project, though consideration of the

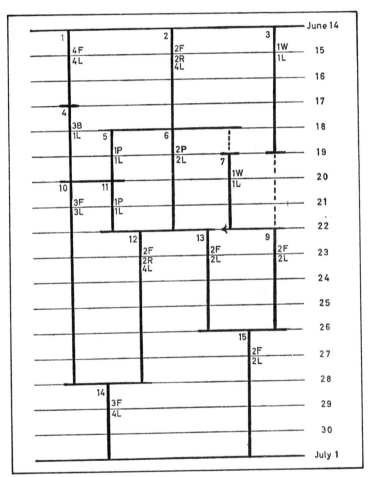

Figure 7.2 Work chart

54

demands made by other projects indicates that the welder can be made available for the first nine days of this project if he is wanted for that time. Fortunately, the job chart shows that each of Jobs 7 and 9 has 1-day float, so that the start of Job 7 can be delayed for 1 day until Job 3 is complete and the start of Job 9 is delayed by 1 day; the one welder completes Job 3 and then moves to Job 7. The amended job chart becomes the work chart of Figure 7.2, in which every job of the project is critical; although the work chart shows some float following Job 3 and preceding Job 7, the labour schedule has eliminated these floats. The corresponding amendments must be made to the manpower array for days 5 to 12; note that the welder is required for the first eight days of the project.

The one-day delay to Job 7 shown in the work chart means that the time line relating the completion of Jobs 3 and 7 to the start of succeeding jobs now occurs at the same point in time as the time line relating the completion of Jobs 6 and 11. It is not permissible to draw one time line here as this would indicate that each of Jobs 9, 12 and 13 could be started only when all of Jobs 3, 6, 7 and 11 are completed. Jobs 12 and 13 are dependent on Jobs 3, 6, 7 and 11, but there is no dependence between Jobs 6 and 11 and Job 9. If it is thought to be useful or informative an arrow can be drawn on the time line at the end of Job 7 to show that a one-way dependency exists; if used, this arrow is of course identical with the dummy arrow used in the arrow network of Figure 4.5. However, the work chart is not intended as an aid in the planning of a project — the planning has been completed — and there is no need to worry about the representation of special relationships in this chart. The work chart is prepared for the sole purpose of providing the project supervisors with a detailed schedule of when the jobs are to be started and completed and of what manpower or other resource is allocated for each job.

Thus the planning, analysis, and scheduling of the project are complete. The anticipated starting date of the project is agreed with plant supervision and dates are written on the work chart. A copy of the work chart is given to every supervisor and foreman concerned with the project and the work chart enables the project supervisors to keep tight control of the project and ensure that no job is delayed or prolonged for any reason whatsoever.

Chapter 8

Job Charts for Multiproject Scheduling

Project analysis consists of three interdependent phases:
1 Planning.
2 Analysis and scheduling.
3 Controlling.
These have been dealt with in detail in Chapter 1 where the construction and use of job charts was described with reference to an example project. Further examples will be given in later chapters, but in this chapter we shall concern ourselves with some aspects of the second phase — that of analysis and scheduling — and show how job charts may be used for determining minimum cost and minimum time projects in the planning stage, how the available resources may be applied to the projects in the scheduling stage, and how the final work charts may be used to control the project when work starts. For this, three projects which are to be run concurrently by the same engineering team will be used.

The maintenance section of a factory is required to complete three engineering projects, with its normal resources, as quickly as possible; the projects are in separate plants but under the control of one engineering manager. The planners produced the job sheets as given in Figures 8.1, 8.3 and 8.5, and from these drew the job charts as shown in Figures 8.2, 8.4 and 8.6. For each project two charts were drawn: a 'standard working' chart which assumes that each job is started at its earliest possible start time and a 'crash working' chart which shows all the jobs crashed that make a contribution to a reduction in the overall project duration. The procedure of developing the crash chart may be worth discussion, and the planners will be followed in detail, Obviously, it is necessary to consider only those jobs which are critical or become critical on reduction of the duration of a critical job. Reduction of the duration of a non-critical job, which remains non-critical whatever happens to any other job, cannot effect any reduction in the overall project duration and is wasteful. Now the standard charts are used.

Job number	Job sequence	Standard working			Crash working		
		Duration (days)	Cost (£)	Labour	Duration (days)	Cost (£)	Labour
1	0;1	3	500	3F	2	700	5F
2	0;2	4	800	2F, 2R	2	1,200	4F, 4R
3	0;3	3	700	3F	2	900	5F
4	1, 2;4	4	950	2F	2	1,500	4F
5	1, 2;5	3	250	1P	2	350	2P
6	1, 2;6	4	400	2F, 2R	2	750	4F, 4R
7	4;7	2	350	2F, 2W	2	350	2F, 2W
8	5, 6;8	5	700	2F	3	900	4F
9	7;9	2	300	2F	2	300	2F
10	9;10	2	400	2F	1	700	4F
11	7;11	8	1,000	2F	5	1,350	3F
12	8;12	6	1,250	3F	4	1,750	5F
13	8;13	4	750	1F	2	1,000	2F
14	8;14	6	1,400	2P	4	2,000	3P
15	10, 12;15	2	350	2F	2	350	2F
16	13;16	4	650	1F	2	900	2F
17	16;17	4	800	1F	2	1,100	2F
18	17;18	2	400	2E	2	400	2E
19	0;19	3	500	4F	2	700	6F
20	19;20	4	550	2F, 2W	2	850	4F, 4W
21	20;21	2	350	2F, 2W	2	350	2F, 2W
22	3, 21;22	5	1,350	2F, 2R	3	1,600	3F, 3R
23	22;23	3	550	2W	2	800	3W
24	11, 14, 15, 18, 23;24	4	1,150	2F	3	1,550	3F
25	11, 14, 15, 18, 23;25	6	1,300	2F	3	1,950	4F
			17,700				

Figure 8.1 Job sheet for Project 1: joint control project

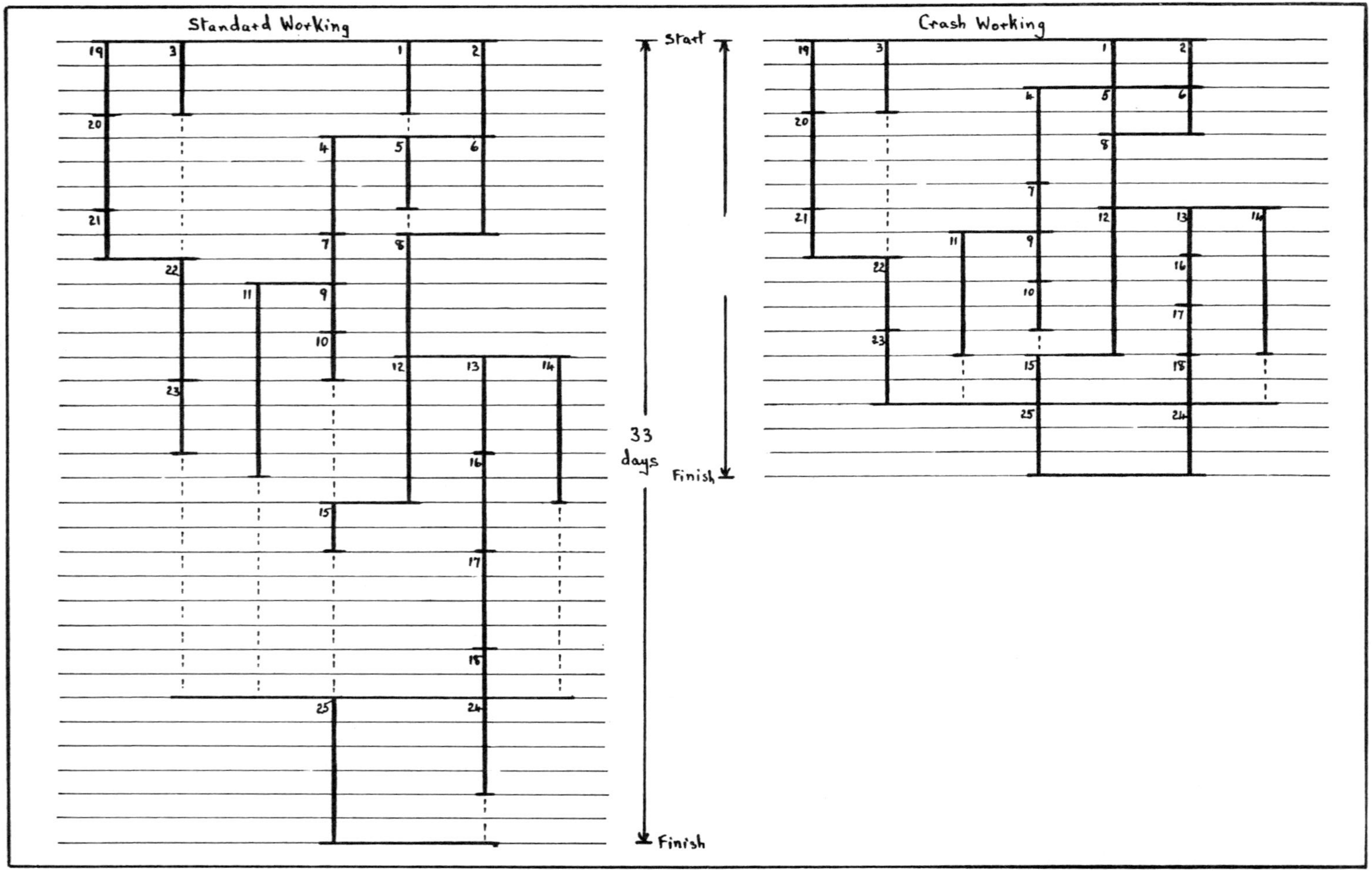

Figure 8.2   Job charts for Project 1

Job number	Job sequence	Standard working			Crash working		
		Duration (days)	Cost (£)	Labour	Duration (days)	Cost (£)	Labour
1	0;1	5	1,130	2F	3	1,450	3F
2	1;2	3	600	2F	3	600	2F
3	2;3	2	450	2F	2	450	2F
4	0;4	2	520	2B	1	750	4B
5	4;5	2	475	2B	1	700	4B
6	0;6	6	1,000	2B	3	1,400	4B
7	0;7	4	800	1P	4	800	1P
8	7;8	4	850	1P	4	850	1P
9	3, 5, 6, 8;9	3	500	2F, 2R	2	650	3F, 3R
10	9;10	2	450	2F	2	450	2F
11	3, 5, 6, 8;11	4	850	4F	2	1,050	8F
12	3, 5, 6, 8;12	2	400	2F	2	400	2F
13	12;13	2	450	2F	2	450	2F
14	13;14	2	420	2F	2	420	2F
15	10;15	4	1,600	2F	3	2,000	3F
16	10;16	3	600	2F	2	750	3F
17	16;17	4	850	2F	2	1,050	4F
18	11, 14;18	7	1,550	2F	4	2,100	4F
19	11, 14;19	3	500	2W	2	650	3W
20	19;20	2	400	2W	2	400	2W
21	15, 18;21	3	550	2F	2	750	3F
22	15, 18;22	6	900	4F	4	1,400	6F
23	17, 21;23	2	350	4F	2	350	4F
24	0;24	16	3,850	2E	10	5,200	3E
25	24;25	2	400	2F	1	650	4F
26	20, 25;26	5	1,100	2E	3	1,600	3E
27	22, 23, 26;27	3	800	4F	2	900	6F
			22,345				

Figure 8.3 Job sheet for Project 2: joint control project

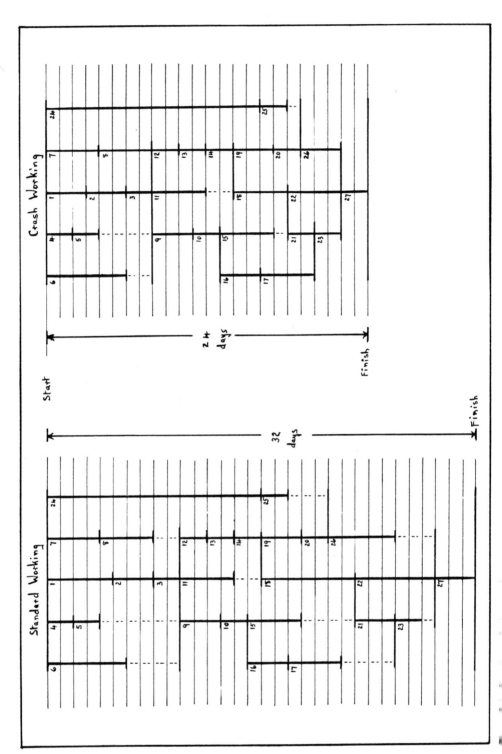

Standard Working

Start

32 days

Finish

Crash Working

24 days

Finish

Job number	Job sequence	Standard Working			Crash working		
		Duration (days)	Cost (£)	Labour	Duration (days)	Cost (£)	Labour
1	0;1	2	200	2C	1	250	4C
2	0;2	2	200	2C	1	250	4C
3	0;3	2	250	2C	1	300	4C
4	0;4	2	250	2C	1	300	4C
5	0;5	2	200	2C	1	250	4C
6	0;6	7	1,500	2F	4	1,950	4F
7	6;7	3	575	2F	2	700	3F
8	0;8	4	650	2F	2	1,000	4F
9	8;9	2	300	1P	1	400	2P
10	8;10	2	270	2F	2	270	2F
11	10;11	2	250	3F	1	350	6F
12	8;12	2	275	1F	1	350	2F
13	9, 10, 12;13	2	300	2F	1	400	4F
14	1, 2, 3, 4, 5, 7;14	3	520	2W	2	700	3W
15	1, 2, 3, 4, 5, 7;15	3	550	2W	2	700	3W
16	15;16	4	700	2F	2	950	4F
17	13, 14, 16;17	4	720	2E	2	900	4E
18	13, 14, 16;18	2	300	2E	1	420	4E
19	13, 14, 16;19	2	250	2E	1	400	4E
20	13, 14, 16;20	13	6,450	6F	8	8,250	10F
21	20;21	5	1,050	4F	3	1,450	7F
			15,750				

Figure 8.5 Job sheet for Project 3: joint control project

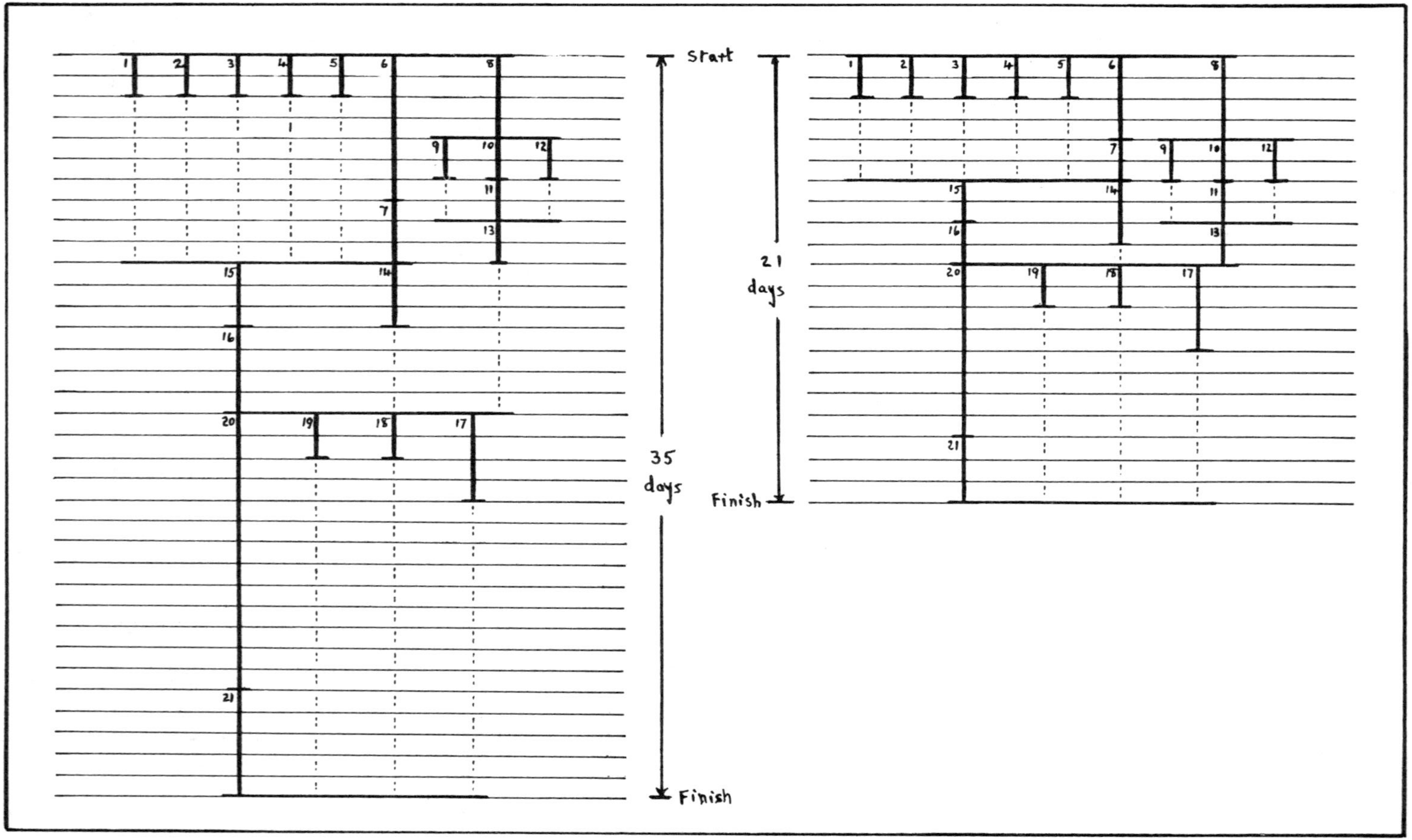

Figure 8.6 Job charts for Project 3

Project 1 (Figures 8.1 and 8.2)

For Project 1, the critical path consists of Jobs 2, 6, 8, 13, 16, 17, 18 and 25. The sequence of Jobs 13, 16, 17 and 18 looks promising, and it can be seen from the job sheet that each of Jobs 13, 16 and 17 can be crashed by 2 days, while Job 18 cannot be crashed. The maximum time saving on this sequence is 6 days, if the saving can be made. Following Job 15 there are 6 days' float, and following Job 14 there are 8 days' float, and no other job is involved. Then we crash the sequence of Jobs 13, 16, 17 and 18 by 6 days, which eliminates the float after Job 15 so that Jobs 12 and 15 become critical; and the float after Job 14 is reduced from 8 to 2 days. The time line at the end of Job 18 has been lifted by 6 days, and so the float after Job 11 is reduced from 9 to 3 days, and the float after Job 23 is reduced from 10 to 4 days.

Next it is seen that Job 8 can be reduced from 5 to 3 days without affecting any other job, but this reduces the float after Job 10 from 5 to 3 days, reduces the float after Job 11 from 3 days to 1 day, and reduces the float after Job 23 from 4 to 2 days. If Job 25 is reduced from 6 to 3 days, Job 24 must be reduced from 4 to 3 days or the full time saving cannot be made. No other job or float is affected here. Similarly, if Job 2 is reduced from 4 to 2 days, Job 1 must be reduced from 3 to 2 days. No other job is affected, but the float after Job 23 is reduced from 2 days to zero.

Only Job 6 and the jobs in the sequence Jobs 19, 20, 21, 22 and 23 (which have become critical) are left. If Job 6 is reduced by 2 days, then Job 5 must be reduced from 3 to 2 days. This reduces the float after Job 10 from 3 days to 1 day and reduces the float after Job 11 to –1 day, which means that Job 11 must be crashed from 8 days to 5 days, restoring the float after Job 11 to 2 days; and these 2 days must be saved in the sequence of Jobs 19 to 23. It is seen that

1   Job 19 can be reduced by 1 day at a premium of £200, i.e. at a cheapness rating of £200 per day.
2   Job 20 can be reduced by 2 days at a premium of £300, i.e. at a cheapness rating of £150 per day.
3   Job 21 cannot be crashed.
4   Job 22 can be reduced by 2 days at a premium of £250, i.e. at a cheapness rating of £125 per day.
5   Job 23 can be reduced by 1 day at a premium of £250, i.e. at a cheapness rating of £250 per day.

Thus Job 22 is chosen to be crashed.

No other job should be crashed as the project is now at minimum duration and twenty of the twenty-five jobs are now critical. Note that in this project Job 22 has been crashed with Jobs 5, 6 and 11 and not with Jobs 1 and 2. Jobs 5, 6 and 11 cannot be separated, and the premium cost of crashing these is £800, while the premium cost of crashing Jobs 1 and 2 is £600. Each gives a saving of 2 days, and since it is necessary to evaluate the worth of crashing, each crash is kept at as low a premium as possible. Now a cheapness rating table can be compiled as in Figure 8.7 in which are listed the jobs that must be crashed to effect any reduction in the project duration and determine the premium cost per day saved. These cheapness ratings must be compared with the value of the project to the plant; if the value (the profit to be made on completion of the project) is less than £100 per day, then no job would be

Crash jobs	Days saved	Premium expenditure	Cheapness rating
8	2	£200	£100 per day
13	2	250	125
16	2	250	125
17	2	300	150
1 and 2	2	600	300
24 and 25	3	1,050	350
5, 6, 11 and 22	2	1,050	525
Project	15	£3,700	£246.6

Figure 8.7 Cheapness ratings for Project 1

Project duration	Project cost
33 days	£17,700
31	17,900
29	18,150
27	18,400
25	18,700
23	19,300
20	20,350
18	21,400

Figure 8.8 Project duration — cost tabulation for Project 1

crashed unless the resources were needed on some other project that could carry the premium. If the value were £250 per day, then Jobs 8, 13, 16 and 17 would be crashed, and if the value were £525 per day or more, then all jobs could be crashed. A project cost-duration can be compiled as in Figure 8.8 showing how the cost increases as the duration becomes less. For display purposes the values given in the table may be drawn on a graph as in Figure 6.3.1; if so, intermediate values do not exist.

Project 2 (Figures 8.3 and 8.4)

The critical path for Project 2 consists of Jobs 1, 2, 3, 12, 13, 14, 18, 22 and 27, and it is seen from the job sheet that Jobs 2, 3, 12, 13 and 14 cannot be crashed; only the implications of crashing the other jobs need to be considered.

If Job 1 is reduced from 5 to 3 days, all jobs following the time line at the end of Job 3 are moved up 2 days. No other job is involved, but the float after Job 5 is reduced from 6 to 4 days; the float after Job 6 is reduced from 4 to 2 days. The float after Job 8 is eliminated so that Jobs 7 and 8 become critical, and the float after Job 24 is reduced from 3 days to 1 day. Job 27 stands on its own and may be reduced from 3 to 2 days without involving any other job or

Crash jobs	Days saved	Premium expenditure	Cheapness rating
27	1	£100	£100 per day
1	2	320	160
18	3	550	183
21, 22 and 26	2	1,200	600
For the Project	8	£2,170	£271 per day

Figure 8.9 Cheapness ratings for Project 2

Project duration	Project cost
32 days	£22,345
31	22,445
29	22,765
26	23,315
24	24,515

Figure 8.10 Project duration — cost tabulation for Project 2

float. Jobs 18 and 22 are left. The job sheet shows that Job 18 can be reduced by 3 days for a premium expenditure of £550, i.e. a cheapness rating of £183 per day, while Job 22 can be reduced by 2 days for a premium expenditure of £500, a cheapness rating of £250 per day. Thus preference is given to Job 18. Job 18 can be reduced from 7 to 4 days without involving any other job, and all jobs following Job 18 move up 3 days. The float after Job 15 is reduced from 4 days to 1 day; the float after Job 17 is reduced from 4 days to 1 day; and the float after Job 26 is eliminated. Thus Jobs 19, 20 and 26 become critical.

Now it is seen that if Job 22 is to be reduced from 6 to 4 days 2 days must be cut from the sequence of Jobs 19, 20 and 26. Since Job 19 can be crashed only by 1 day, Job 20 cannot be crashed. Job 26 can be crashed by 2 days; obviously Job 26 must be crashed to provide the 2 days. Also 2 days must be cut from the sequence of Jobs 21 and 23, and here only Job 21 can be crashed, by 1 day, which eliminates the float after Job 17 so that Jobs 16 and 17 become critical. The float after Job 23 is eliminated so that Jobs 21 and 23 become critical. Now the cheapness rating tabulation is compiled as in Figure 8.9 and the project cost-duration table as in Figure 8.10. Note that, if desired, alternative values such as those obtained by crashing Job 22 before Job 18, or Job 18 before Job 22, may be included in the project cost-duration table but should not be included in the cheapness rating table.

Project 3 (Figures 8.5 and 8.6)

The critical path for Project 3 consists of Jobs 6, 7, 15, 16, 20 and 21 and the crash programme is obvious. Job 6 can be crashed from 7 to 4 days. No other job is involved, but the time line at the bottom of Job 7 is raised by 3 days. The float after Job 13 is reduced from 7 to 4 days; each of the floats after Jobs 1, 2, 3, 4 and 5 is reduced from 8 to 5 days.

Job 7 can be crashed from 3 to 2 days. No other job is involved, but the time line at the bottom of Job 7 is raised by 1 day. The float after Job 13 is reduced from 4 to 3 days; each of the floats after Jobs 1, 2, 3, 4 and 5 is reduced from 5 to 4 days.

Job 15 can be reduced from 3 to 2 days. No other job is involved, but the time line at the bottom of Job 16 is raised by 1 day. The float after Job 13 is reduced from 3 to 2 days, and the float after Job 14 is reduced from 4 to 3 days. Job 16 can be reduced from 4 to 2 days. No other job is involved, but the time line at the bottom of Job 16 is raised by 2 days. The float after Job 14 is reduced from 3 days to 1 day, and the float after Job 13 is eliminated so that Jobs 8, 10, 11 and 13 become critical.

Job 20 can be reduced from 13 to 8 days, and Job 21 from 5 to 3 days. No other job is involved, but the float after Job 17 is reduced from 14 to 7 days, and each of the floats after Jobs 18 and 19 is reduced from 16 to 9 days.

Now the cheapness rating tabulation can be compiled as in Figure 8.11 and the project cost-duration tabulation as in Figure 8.12. Some of the jobs have equal ratings and the cost-duration tabulation must be written in full.

The initial planning stage is complete. Job charts for minimum cost and for minimum time have been prepared, and the projects have been costed for

Crash jobs	Days saved	Premium expenditure	Cheapness rating
7	1	£125	£125 per day
16	2	250	125
15	1	150	150
6	3	450	150
21	2	400	200
20	5	1,800	360
For the Project	14	£3,175	£227

Figure 8.11 Cheapness ratings for Project 3

Project duration	Project cost	Jobs crashed
35 days	£15,760	
34	15,885	7
33	16,010	16
32	16,135	7 and 16
31	16,285	7, 16 and 15
29	16,585	7, 16 and 6
28	16,735	7, 16, 15 and 6
26	17,135	7, 16, 15, 6 and 21
20	18,935	7, 16, 15, 6, 21 and 20

Figure 8.12 Project duration — cost tabulation for Project 3

feasible durations. Now it is necessary to schedule the jobs with respect to the available resources and determine a feasible work programme to cover the three projects.

Scheduling the projects

The Works Manager instructs that Project 1 is to be completed in 25 days for £18,700 and that all three projects are to be completed in 40 days. No premium expenditure is to be incurred in Projects 2 and 3 unless necessary to complete the work within the time limit. This means that in Project 1, Jobs 8, 13, 16 and 17 are to be crashed, and scheduling the three projects without crashing any other job is to be attempted. The limits on resources are fixed at

2	Bricklayers	15	Fitters	2	Riggers
2	Carpenters	2	Plumbers	2	Welders
2	Electricians				

Any of these not wanted on the projects can be utilised elsewhere. Further, the restriction is applied that no job may be divided into smaller parts, though very often it is found that many jobs may be divided without interfering with other jobs. The first step is to see how these resources are likely to restrict continuity of work. So the three project charts are drawn on one sheet as in Figure 8.13, starting each from the same start time, as there is no reason to think as yet that the projects cannot be started at the same time. The chart for Project 1 shows the 25-day project, while the charts for Projects 2 and 3 are the standard working charts for these projects. Next the manpower requirements are entered on each jobline; the convention has been adopted that job numbers will be written to the left of the job line, and resource requirements will be written to the right of the job line. Now the manpower requirements are summed across the three charts and the daily total requirements are entered in the manpower array at the right-hand side of Figure 8.13.

Immediately it is seen that Jobs 4, 5 and 6 in Project 2 must be sequenced, and Jobs 1, 2, 3, 4 and 5 in Project 3 must also be sequenced; available float permits these sequences to be made without affecting any other jobs, and this satisfies requirements for bricklayers and carpenters. The 3-fitter excess shown in Figure 8.13 for Days 1, 2 and 3 can be eliminated by delaying the start of Project 1 Job 3 for 3 days, and the schedule chart for Project 1 is begun, as in Figure 8.14, by drawing Job lines 1, 2, 3, 4, 5, 6 and 8, noting that Job 3 can be delayed a further 3 days if required. It is noted that both Jobs 7 and 21 need 2 welders, and these two jobs have a 1-day overlap on Figure 8.13. The situation is rectified by delaying the start of Job 7 by 1 day, which eliminates the float after Job 11, reduces the float after Job 10 from 3 to 2 days, and allows a 1-day float after Job 4. The other job lines are drawn in Figure 8.14 as in Figure 8.13.

Now the schedule charts can be compiled for Projects 2 and 3, and for brevity the notation A-X will be used, in which X refers to a job number in Project A so that, for example, 3-21 means Project 3 Job 21. Straight away the job lines can be drawn for 2-1, 2-2, 2-3, 2-4, 2-5, 2-6, 2-7, 2-8 and 2-24, and the job lines for 3-1, 3-2, 3-3, 3-4, 3-5, 3-6, 3-7 and 3-8. Before going any further the resources which enforce the most severe restrictions — riggers,

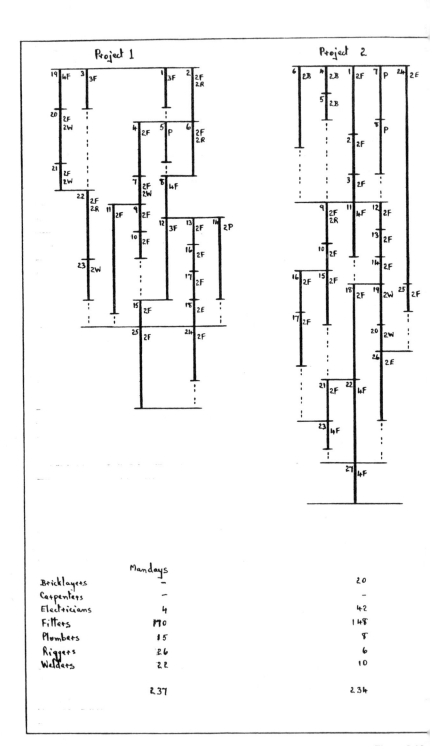

	Mandays	
Bricklayers	—	20
Carpenters	—	—
Electricians	4	42
Fitters	170	148
Plumbers	15	8
Riggers	26	6
Welders	22	10
	237	234

Figure 8.13

Project 3

B	C	E	F	P	R	W	Day
4	10	2	15	1	2	–	1
4	10	2	15	1	2	–	2
4		2	15	1	2	–	3
4		2	10	1	2	2	4
2		2	14	3	2	2	5
2		2	14	3	2	2	6
		2	13	2	2	2	7
		2	13	1	2	2	8
		2	14	–	–	4	9
		2	16	–	2	2	10
		2	18	–	4	4	11
		2	19	2	4	4	12
		2	19	2	4	4	13
		2	21	2	2	–	14
		2	13	2		2	15
		2	15	2		2	16
		–	17	2		4	17
		8	18			2	18
		8	14			2	19
		2	14			2	20
		2	14			2	21
		2	12				22
		2	12				23
		2	14				24
		2	14				25
		2	12				26
			14				27
			14				28
			10				29
			10				30
			8				31
			8				32
			4				33
			4				34
			4				35

B	C	E	F	P	R	W
–		20				
20		20				
16			62			
152			470			
2				25		
–					32	
12						44
202						673 mandays

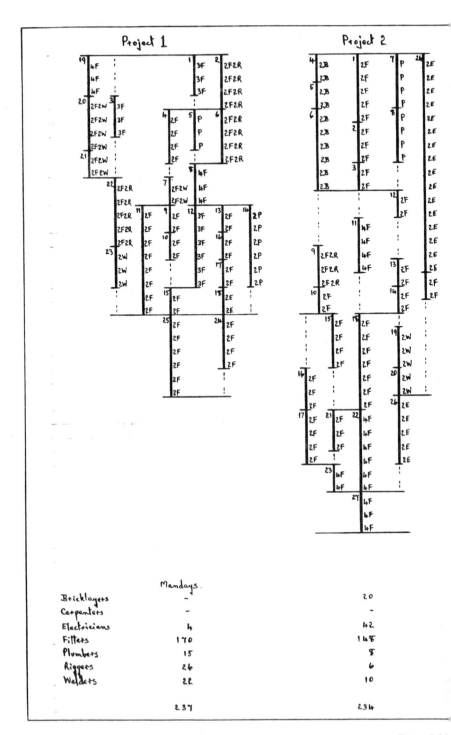

Figure 8.14

·ject 3

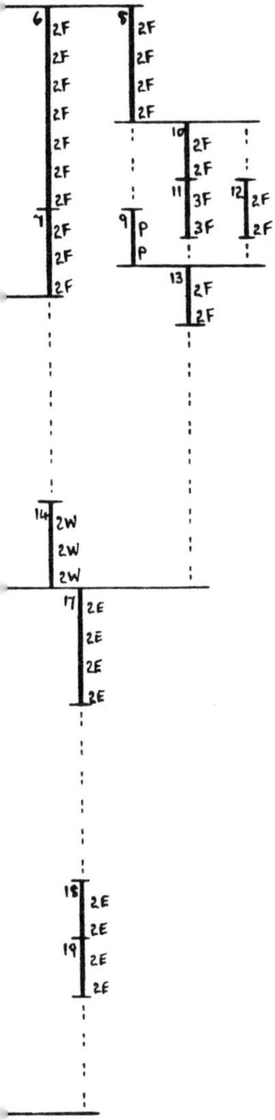

Diagram (left column):

```
6 |2F      5 |2F
  |2F        |2F
  |2F        |2F
  |2F        |2F
  |2F     10 |2F
  |2F        |2F
  |2F     11 |3F   12 |2F
7 |2F    9 |P |3F    |2F
  |2F      |P
  |2F     13 |2F
  |2F        |2F

14 |2W
   |2W
   |2W
17 |2E
   |2E
   |2E
   |2E

18 |2E
   |2E
19 |2E
   |2E
```

B	C	E	F	P	R	W	Date
2	2	2	15	1	2	–	Sept. 5, 1966
2	2	2	15	1	2	–	6
2	2	2	16	1	2	–	7
2	2	2	13	1	2	2	8
2	2	2	15	2	2	2	9
2	2	2	15	2	2	2	10
2	2	2	15	2	2	2	11
2	2	2	15	2	2	2	12
2	2	2	10	1	–	2	13
2	2	2	14	–	2	2	14
		2	12	–	2	2	15
		2	13	2	2	2	16
		2	15	2	2	2	17
		2	15	2	2	2	18
		2	15	2	2	2	19
		2	15	2	2	2	20
		–	15	2	2	2	21
		2	12			2	22
		2	10			2	23
		–	10			2	24
		2	14			2	25
		2	14			2	26
		2	14			2	27
		2	12			2	28
		–	12			2	29
		2	10				30
		2	14				Oct. 1
		2	14				2
		2	14				3
		2	12				4
		2	14				5
		2	14				6
		2	10				7
		2	8				8
			8				9
			4				10
			4				11
			4				12

Totals:

B	C	E	F	P	R	W
–		20				
20	20					
16			62			
152			470			
2				25		
–					32	
12						44
202			673			

gements of projects

71

welders and plumbers — must be considered. At this stage it is not necessary to be too concerned with the requirements for 8 electricians on Days 18 and 19, as there appears sufficient float after 3-17, 3-18 and 3-19 to sequence the jobs needing electricians without much difficulty. It is seen that 2-9 cannot start until Day 15, after 1-22 has been completed; and 3-9 cannot start until Day 8, after 1-5 has been completed. These two jobs fill all riggers and plumbers requirements and may be drawn in Figure 8-14, and the manpower array is written. It seems likely that 3-14 and 3-15 will be required before 2-19 and 2-20, and so 3-15 for Days 12, 13 and 14 is entered; 3-14 for Days 18, 19 and 20; then 2-19 is drawn for Days 21, 22 and 23 and 2-20 for Days 24 and 25. Tentatively, 3-17 is placed in Days 21, 22, 23 and 24, noting that this position depends on how 3-16 may be placed; 2-26 in Days 26, 27, 28, 29 and 30; 3-18 in Days 31 and 32; and 3-19 in Days 33 and 34. These are all subject to the ability to fit in the remaining jobs under the 15-fitter restriction.

So 3-10 is placed in Days 5 and 6 (there is no reason why 3-10 should not be placed before 3-9 as there is no dependence between these two jobs); 3-11 and 3-12 are placed in Days 7 and 8; and 3-13 is placed in Days 10 and 11. The remaining jobs are placed to meet the 15-fitter restriction, and it is seen that 2-12 can be started on Day 11; 2-11 cannot be started until Day 13; 2-13 cannot be started until Day 16; 3-16 cannot be started until Day 17; and so on. As they now stand the projects are

Project 1 is completed in 25 days at a cost of £18,700.
Project 2 is completed in 35 days at a cost of £22,345.
Project 3 is completed in 38 days at a cost of £15,760.

All requirements and restrictions have been met. Note that the resource restrictions are such that even if premium expenditures are permitted on Projects 2 and 3, the only reductions in job durations that can be made are that 2-27 can be crashed from 3 to 2 days, and 3-21 can be crashed from 5 to 3 days. Feasible work charts are available. Figure 8.14 is issued to all personnel concerned, and the scheduling stage is complete. Note that it does not matter how many projects are to be scheduled; as long as a priority is established the procedure is simple with the aid of job charts.

Controlling the projects

A starting date is agreed and the project controller allocates workmen to the individual jobs, issuing any special instructions and briefing the foremen who are to control the work. Each foreman is given a copy of the work charts as on Figure 8.14, and on these charts the progress of the work is recorded. Thus tight control can be maintained over all the work in hand; any improvement or reduction in work speed from that envisaged in the job sheet is noted as soon as it occurs. Whereas a reduced work speed may necessitate some additional help on an emergency basis, an improved work speed will be reflected in future plans.

It has been shown how job charts can be used in all three stages of maintenance, replacement and installation work in a factory, and there can be no doubt that these charts simplify all the work involved.

Chapter 9

Job Charts for Projects from the Monsanto Works

Most of the projects that occur in a chemical works are small, each consisting of only a few job elements, usually up to thirty or forty jobs in a project. Even so the job charts for these projects are well worth the time spent in their construction, and in this chapter some of these projects will be considered. Occasionally, larger projects arise and some examples of these are dealt with in Chapter 10. All these projects are planned, scheduled and controlled by means of job charts. The works engineers and foremen draw the charts for the projects under their control and circulate copies to all the supervisors concerned. As might be expected, most of the projects deal with repair or installation of chemical manufacturing equipment and plant supervision is vitally concerned. Some of the projects, however, are purely workshop projects that can be carried out as and when labour may be made available, and these are the concern of engineering staff only.

The projects can be classified as:

1 Routine repetitive and frequent projects such as cleaning and overhauling mills; overhauling centrifuges; repairing furnaces; renewing pipelines; overhauling or changing reaction vessels, pumps, motors, agitators, fans, etc. Such projects as these occur many times each month. Often they result from breakdown of equipment which demands immediate attention. Job charts facilitate maximum utilisation of resources and lead to minimum down-time of equipment; in some cases the job chart has become a standard.

2 Routine but infrequent projects such as plant overhauls and plant modifications. Usually these projects are planned and scheduled well in advance of the plant closures. The job chart is used for the planning and scheduling of the project, and to control the work once this is started.

3 Minor installation projects such as plant replacements or extensions. Such projects as these, which involve the outlay of additional capital in a plant, are considered, discussed, argued over, planned and scheduled well in advance of the start of work on the plant. Extension projects may not involve closure of the plant concerned, but may need very careful planning and scheduling to ensure that the work flow agreed on the final plan can be performed without hindrance to the continuing manufacturing operations. Use of a job chart enables the project to be handled through all its stages with the minimum of fuss and bother.

4 Special rehabilitation projects such as the relocation of manufacturing equipment or centralisation of workshop facilities. Projects such as these need extensive planning and scheduling to prevent general confusion and dislocation when the work is being carried out. The planning and scheduling are made much easier by the use of job charts.

5 Workshop projects such as retubing a condenser; stripping and reassembling a pump; repair and testing of a temperature recorder. These projects do not interfere with plant operation as they are performed on a unit which has been removed from a plant; the removal will be a job in some other project. The project may be carried out piecemeal as labour becomes available. Job charts facilitate control of the work, and offer an easy method of checking progress at any time.

Examples of each of these projects are considered in detail, using the job sheets and job charts prepared by the works engineers. As far as is known, arrow networks have never been drawn for these projects or any other projects undertaken at the Monsanto works, and none is drawn here. If the reader prefers to use arrow networks, the examples may provide him with a little practice. In the next chapter the uses of arrow networks and job charts for large projects are compared.

Project 4 — Routine and frequent project — Cleaning and overhauling a mill

Very few chemical manufacturing plants have installed spare equipment, and it is not customary to provide any plant with a spare mill. However, in most cases the equipment is designed so that the mill throughput is greater than the manufacturing capacity of the plant, accepting that it will be necessary to stop the mill regularly and frequently for cleaning and overhaul.

Whenever a mill is shut down, the plant output is reduced (or maybe stopped altogether if the mill is the only unit available) while manufacture continues. Thus results in an accumulation of product ready for milling, and it can be readily understood that a prolonged stoppage at the mull may result in too great an accumulation of mill feed, thereby forcing a shut-down of the manufacturing units. And even small losses of production cannot be tolerated on a regular basis. Therefore it is necessary to ensure that a mill is out of action for the minimum time required to do the job that is to be done. This means that the right men must be in the right place at the right time to get the right job done as smoothly and quickly as possible. Preparation of a job chart helps in the scheduling of work and men. The job sheet for the project is given in Figure 9.1 and lists all the jobs of the project with the manpower

Job number	Job	Job sequence	Job time (min.)	Labour requirement
1	Rig up lifting gear	0;1	15	R
2	Clean all equipment with vacuum cleaner	0;2	15	P
3	Remove fan ducting	1, 2;3	30	R, F
4	Remove cyclone separator	3;4	25	R, F
5	Clean ducting	3;5	30	P
6	Clean cyclone separator	4;6	60	P
7	Dismantle and remove beater unit, etc.	4;7	35	F
8	Remove final cone	7;8	10	R, F
9	Clean final cone	8;9	60	P
10	Remove further ducting	8;10	10	R, F
11	Clean further ducting	10;11	20	P
12	Inspect and adjust beater unit	7;12	25	F
13	Inspect and adjust cyclone separator	6;13	15	F
14	Replace further ducting	11;14	10	R, F
15	Replace final cone	9-14;15	10	R, F
16	Replace cyclone separator	15;16	15	R, F
17	Replace fan ducting	5, 16;17	30	R, F
18	Replace beater unit, etc.	12, 17;18	20	F
19	Remove lifting gear	14-18;19	15	R
20	Clean up	All;20	30	P, F

F = Fitter. P = Process Operator. R = Rigger.

Figure 9.1 Job sheet for Project 4: cleaning and overhauling a mill

requirements for each job. Figure 9.2.1 shows the job chart for the project, constructed in the usual first feasible plan method with all floats following the floaters. This chart shows that twelve of the twenty jobs in the project form the critical path, and also shows that some of the jobs may be carried out concurrently. Jobs 8 and 12, and then Jobs 10 and 12, may be carried out concurrently if two fitters can be allocated to the project. Jobs 6, 9 and 11 may be carried out concurrently if three teams of process operators are available. Since this project is to be completed with the services of only one fitter, and since there are only two teams of process operators who can be released from other duties to help out with this project, it is necessary to provide a job chart which meets these restrictions. The position in time, as shown in the original job chart, of some of these jobs must be moved, using the floats and splitting these floats in any feasible manner so that all the jobs can be accommodated with the manpower laid down. If possible, this is to be done without extending the project duration.

The positions of four of the jobs need to be moved, and Figure 9.2.2 shows the final work chart for the project. On completion of Job 8 the fitter is required to attend to Jobs 10, 12, 13 and 14 in that order, and these jobs form a branch critical path. One team of process operators attends to Jobs 5 and 9,

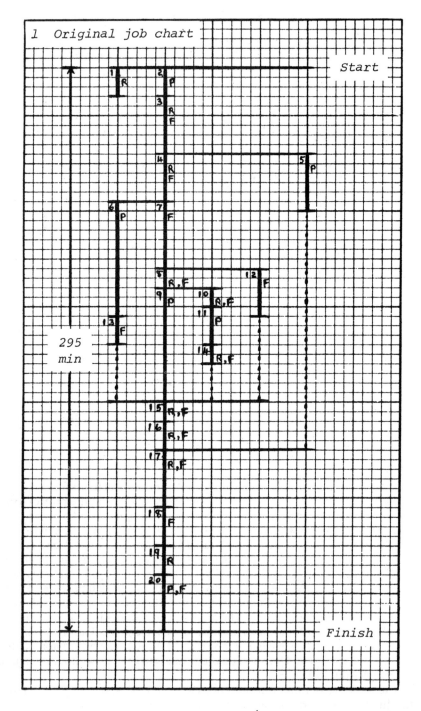

Figure 9.2

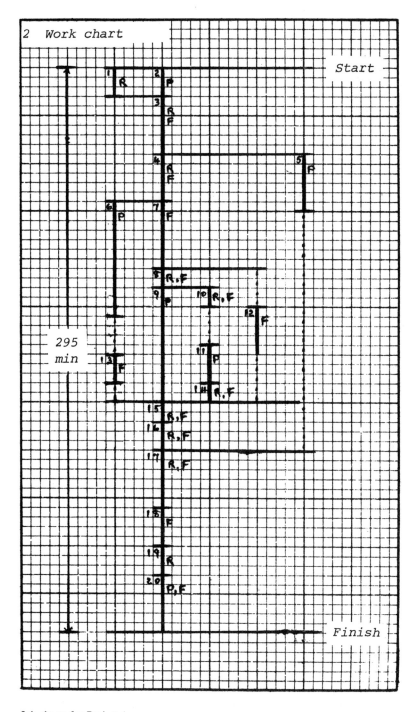

Job charts for Project 4

while the other team attends to Jobs 6 and 11. The project duration remains at 295 minutes, and Jobs 5 and 6 are the only two jobs in the project to be followed by float; after Job 5 the process operators have 30 minutes before they have to start on Job 9, and after Job 6 the other team of process operators has 15 minutes before Job 11 must be started. As far as these jobs are concerned it is just as correct to relate the float to the teams of operators as to the jobs.

This project is a very small example of the way in which scheduling can impose restrictions on a project and create a series of critical jobs.

Project 5 — Routine but infrequent project — Modifications to fan and dryer

At some stage or another in a project such as this, complete closure of the plant, with its consequent loss of production, will be involved. The date for shutting down the plant is decided by the production manager, who will be advised by the engineer on forward dates for delivery of equipment and on the expected duration of the shut-down. When the date of the plant shut-down is agreed, the work must be planned and scheduled so that all the preparatory work is completed by the shut-down date; once the plant is shut-down no delay can be permitted and manufacture must be restarted as soon as possible.

Figure 9.3 gives the job sheet for this project; the jobs are detailed and their manpower requirement quoted. If we spend a moment considering some of the jobs we see that two of them, Jobs 3 and 8, relate to delivery of equipment. When the project is authorised and the equipment ordered, a tentative delivery date is quoted. Quite often this delivery date gets amended, maybe several times, but eventually a firm delivery date is obtained and the project becomes 'live' from this date. This date is taken as the starting point for construction of the job chart. Job 1, too, has been in hand since the project was authorised, and work on this job has been carried out intermittently as and when a welder has been available. Then the job chart for the preparatory work might be as shown in Figure 9.4. Note from this chart that although the work can be completed in $3\frac{1}{2}$ days if manpower is made available as required, the work involves 5 days' work for a fitter; and it is convenient to arrange a work schedule for 5 days so that only one fitter is engaged on the project. Note also that Job 10 can be split into two parts: one part can be started as soon as Job 3 is complete, but the second part cannot be started until both Jobs 3 and 9 are complete. (Job 10 could have been divided into two jobs when the job sheet was compiled). There are several ways in which a 5-day schedule can be arranged for the fitter, as for example by splitting Job 10

1st day	:	Job 2
2nd day	:	First part of Job 10
3rd day	:	Job 5
4th day	:	Second part of Job 10
5th day	:	Job 11

This is the schedule adopted in the work chart shown in Figure 9.5. If for any reason it becomes convenient, or necessary, to run Job 10 as a single job without splitting it, a possible schedule would be

Job number	Job	Job sequence	Job time (hr.)	Labour requirement
1	Fabricate duct for fan	0;1	16	W
2	Install fan packers	0;2	8	F
3	Delivery of base plate	0;3	8	— (S)
4	Inspect and check measurements of base plate	3;4	8	I
5	Install fan and base plate	2, 4;5	8	F
6	Install duct for fan	1, 5;6	8	W
7	Wire up fan motor	5;7	8	E
8	Delivery of equipment	0;8	8	— (S)
9	Inspect and check dimensions of equipment	8;9	12	I
10	Prepare supporting steel-work	3, 9;10	8+8	F
11	Check and grease existing bolts for easy removal	0;11	8	F
12	Erect scaffolding	0;12	8	R
13	Remove air filter and adjacent ducting	10, 11, 12;13	4	F, 2R
14	Remove heater duct	13;14	2	F, R
15	Remove hopper and supports	14;15	2	F, R
16	Place new heater duct	15;16	2	R
17	Place new feed mechanism	16;17	2	R
18	Place air filter and ducting	17;18	4	R
19	Place remaining new ducting	18;19	4	R
20	Fix ducting 16	15,16 part;20	4	F
21	Fix mechanism 17	17, 20;21	4	F
22	Fix ducting 18 and 19	18, 19, 21;22	6	F
23	Wire up motor and starter of 17	17;23	6	E
24	Install steam and gauge lines	16;24	14	F
25	Complete welds on ducting	13-18;25	16	W
26	Dismantle scaffolding and clear site	18;26	8	R

E = Electrician; F = Fitter; I = Inspector; R = Rigger;
(S) = Suppliers; W = Welder.

Figure 9.3 Job sheet for Project 5: modifications to fan and dryer

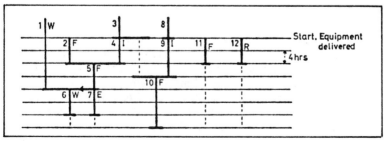

Figure 9.4 Job chart for preparatory work on Project 5

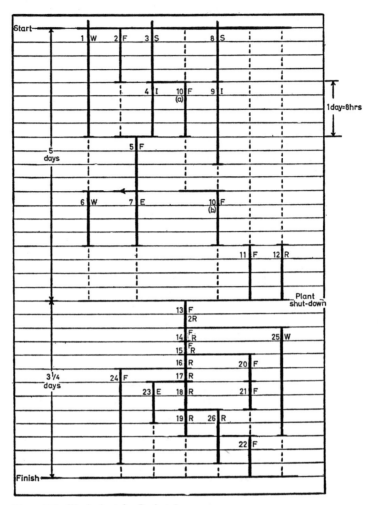

Figure 9.5 Work chart for Project 5

1st day : Job 2
2nd day : Job 11
3rd day : Job 5
4th day : } Job 10
5th day : }

Having scheduled the work to be done by the fitter — which means that for the 5-day schedule these jobs become critical — the timing of the jobs requiring attention from a welder must be considered. Job 6 must be completed in the last 2 days, and Job 1 must be completed in the first 3 days. These two jobs are scheduled to allow a 1-day float after each in case any additional work becomes necessary or some other emergency arises. The work chart shows a direction arrow on the time line at the end of job line 5; this indicates that Job 6 is dependent on both Jobs 1 and 5, but Job 7 is dependent only on Job 5 and has no connection with Job 1. When the plant is shut down, the work is scheduled for completion in minimum time, and the craftsmen required to achieve this minimum time are made available as and when they are needed. Figure 9.5 shows that Jobs 13, 14, 15, 16, 17, 18, 19 and 22 are critical, but Jobs 24 and 26 have only 2 hours' float, and could become critical very easily.

From the job sheet it is seen that Job 25 follows after the sequence of Jobs 13-18. But Job 25 can be started as soon as Job 13 is finished and continues in sections as each of the other jobs is finished. In the work chart, Figure 9.5, Job 25 could be represented by a series of job lines, each following its leader from the sequence of Jobs 13-18, and this would be done if there were any differences of importance in the resource requirements for the sections of Job 25. However, in this case the welder starts work as soon as Job 13 is finished and moves along completing the duct welding as he can. It is more convenient, as far as manpower scheduling is concerned, to show this welding work as one job. The same conditions do not apply to Job 22, which follows completion of Jobs 18, 19 and 21. In this particular project it was not possible for the fitter to start work on Job 22 until Jobs 18 and 19 had been completed and the rigger and his tackle had left the site — the welder was working there also. If the fitter had been able to start work on completion of Job 18, and before Job 19 had been completed, then job line 22 would have followed immediately after the time line at the end of job line 18 and would have continued for the estimated time after completion of Job 19. In this latter case the critical path for the plant shut-down part of the project would have consisted of Jobs 13, 14, 15 and 16 branching to Job 24 on the one side and to Jobs 17, 18 and 26 on the other.

Project 6 — Routine infrequent project — Changing a 2,000-gallon reactor

Obviously, a project such as this cannot be started until the replacement reactor is available. Since reactors may develop faults without giving notice, it is customary to keep spare reactors in stock so that a replacement reactor is on hand, however short the notice of its requirement may be. Wherever possible, of course, changing the reactor is planned to be carried out at the same time as other work on the plant or even a general plant overhaul, since the plant will have to be shut-down. Then two cases must be considered: the project carried out under planned normal overhaul conditions, that is, to standard time; and the project rushed as an emergency, that is, to crash time. For projects such as this the job chart has two useful functions: it portrays the project clearly, and it shows which jobs should not be crashed and those which must be crashed.

The job sheet for this project is given in Figure 9.6. The jobs are detailed, and standard time and crash time are quoted for each job. Note that there are six jobs, Jobs 10, 11, 15, 18, 20 and 22, for which the standard time and crash time are the same. The reasons vary. In some cases, such as Jobs 10 and 15, no additional labour can be employed as the working space is full; in other cases, such as Jobs 11 and 20, the work is controlled by equipment or testing requirements. Figure 9.7 shows the job charts for standard and crash time working. In the chart for standard time, Figure 9.7.1, the critical path is seen to consist of Jobs 1, 2, 5, 7, 9, 10, 11, 13, 14, 16, 19 and 22, while in the crash project, Figure 9.7.2, these jobs plus Jobs 15, 17, 20 and 21 are critical. The

Figure 9.6 Job sheet for Project 6: changing a 2,000-gallon reactor

Job numbers	Job	Job sequence	Standard time (hr.)	Crash time (hr.)
1	Erect lifting gear and working platforms	0;1	16	10
2	Disconnect all electrical wiring	1;2	8	6
3	Remove all insulation from reactor and pipework	1;3	14	10
4	Disconnect and remove all connecting pipework	1;4	16	12
5	Remove gearbox and agitator	2;5	12	8
6	Transport gearbox to workshops and overhaul	5;6	24	16
7	Remove outlet box and other fittings	5;7	12	8
8	Remove vessel plinths	7;8	6	4
9	Remove chequer plate and steelwork	7;9	10	8
10	Remove temporary access platform	9;10	4	4
11	Lower reactor on to transport	3, 4, 8, 9, 10;11	6	6
12	Transport reactor to scrap yard and dump	11;12	x	x
13	Lift new reactor on to steelwork	11;13	8	6
14	Line up reactor and replace steelwork and plinths	13;14	12	8
15	Replace access platforms	13;15	8	8
16	Re-install agitator and gearbox and fittings	6, 14, 15;16	16	12
17	Reconnect pipework	15;17	18	12
18	Reconnect all electrical wiring	15;18	8	8
19	Relag reactor and pipework	16, 17;19	16	12
20	Carry out hydraulic tests	16, 17, 18;20	8	8
21	Paint reactor and pipework	20;21	6	4
22	Clear up site	19, 21;22	4	4

critical jobs in the crash project include five of the six fixed-duration jobs. The job chart for standard time shows that there is no point in crashing Jobs 3, 4, 6, 8, 15 and 18; Job 15 remains at its standard time but becomes critical in the crash project, while Jobs 3, 4, 6, 8 and 18 remain non-critical.

No duration is quoted in the job sheet for Job 12 for either standard or crash working. As soon as the old reactor is loaded on to the transport vehicle and taken from the immediate scene of operations, it is no longer of interest to this project. If the vehicle is to be used in a series of projects, then the total duration of its employment in taking away the old reactor from plant to scrap yard will appear on the vehicle schedule, but the only requirement as far as our Project 6 is concerned is that the vehicle should be on the site at the right time to load the old reactor. Job line 12 is drawn on the job chart at any convenient (but short) length so that it may be readily seen, and the job end-

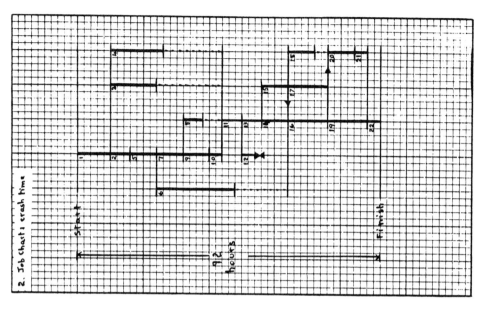

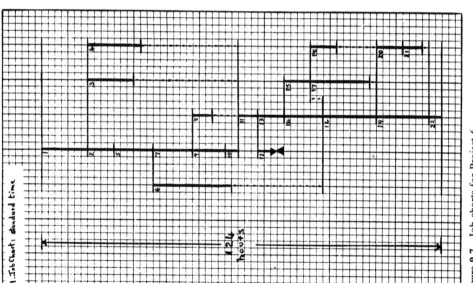

Figure 9.7   Job charts for Project 6

83

line is closed with a gate on some other device to indicate that there is no later job in the project following in sequence from Job 12. Alternatively, it is permissible to hang a circle from the time line at the end of Job 11 and put the number 12 in the circle. In the standard time chart of Figure 9.7.1 note the arrow on the time line following job line 16 which indicates that Job 19 is dependent on the completion of Jobs 16 and 17 only, while Job 20 is dependent on the completion of Jobs 16, 17 and 18. In the crash time chart of Figure 9.7.2 two such arrows are used: one as described for Figure 9.7.1; and the other indicating that Jobs 17 and 18 are dependent on the completion of Job 15 only, while Job 16 is dependent on the completion of Jobs 6, 14 and 15. As has been mentioned previously, these arrows are drawn on the job charts when these charts are used for planning and scheduling, but are not drawn on the charts used as work control charts.

Project 7 — Minor installation project — Installation of new chemical process equipment

Very often during the planning stage of a project it is difficult to decide exactly how much detail is required in the breakdown of the project into its individual jobs. In Project 4 work done on each separate part of the mill was classified as a 'job'; while in Project 5 it was convenient to schedule a lot of work on different sections of plant as one job, because the work was to be completed by one welder moving from one section to the next without interruption, and this arrangement of 'jobs' can be extended. In projects such as Project 7 it has become customary to arrange 'jobs' so that on completion of the work involved in each of these jobs, another major 'stage' in the project is achieved, irrespective of the number of men working on the job or the number of crafts employed. The job sheet, Figure 9.8, shows a number of these collective jobs, and separate break-down work charts for these jobs are prepared by the supervisors concerned as and when required. Thus the job sheet of Figure 9.8 may be called the master job sheet for the project.

As with so many other projects, engineers may be requested to install the new equipment in the shortest possible time so that the additional manufacturing capacity may be made available at the earliest possible moment. Arrangements must be made for crashing the project if this can be shown to be an economic proposition. Since crashing a job requires the use of additional facilities — mostly manpower, and usually at overtime rates of pay — the final cost of the project is likely to show a considerable increase over the cost of normal working. Thus a job is crashed if it satisfies the three restraints

1   That the additional facilities can be made available and can be used on the project without causing undue congestion.
2   That the crashing results in a reduction in the overall project duration.
3   That the cost of crashing is recouped by the additional profit made in the time saved.

From the job sheet, Figure 9.8, it is seen that 19 of the 23 jobs can be crashed. But when these jobs are considered with the normal time job chart of Figure 9.9.1, it is obvious that crashing some of the jobs will not affect the overall project duration in any way. The normal time job chart shows that 9 jobs, Jobs 1, 2, 3, 5, 7, 9, 20, 22 and 23, are critical, and the project duration is 62 days. Jobs 8, 21, 22 and 23 cannot be crashed, and crashing Jobs 4, 10, 11,

Job number	Job	Job sequence	Job time (days) normal	Crash	Crash ?
1	Clear site	0;1	4	2	Yes
2	Excavation and foundations	1;2	9	6	Yes
3	Install support steelwork	2;3	9	7	Yes
4	Install stairways	3;4	3	2	—
5	Install platforms	3;5	8	5	Yes
6	Install handrails and platform angles	3;6	7	5	Yes
7	Install equipment	4, 5, 6;7	4	3	Yes
8	Calibrate equipment	7;8	7	7	—
9	Install prefabricated ductwork	7;9	21	14	Yes
10	Install prefabricated process pipework	4, 5, 6;10	10	7	—
11	Make connections pipework to equipment	8, 10;11	4	2	—
12	Install prefabricated service pipework	4, 5, 6;12	7	6	—
13	Make connections service pipework to equipment	8, 12;13	3	2	—
14	Install instrumentation	4, 5, 6;14	14	7	—
15	Make connections instruments to equipment	8, 14;15	4	2	Yes
16	Install electrical apparatus	4, 5, 6;16	9	6	—
17	Make electrical connections to equipment	8, 16;17	3	2	—
18	Lag equipment	8;18	10	5	—
19	Lag pipework	11, 13;19	14	7	Yes
20	Lag ductwork	9;20	6	3	—
21	Testing equipment	1-17;21	4	4	—
22	Painting	1-13;22	7	7	—
23	Clean up	1-20;23	1	1	—

Figure 9.8 Job sheet for Project 7: installation of new chemical process equipment

12, 13, 14, 16, 17, 18 and 20 will not help. The crash time work chart of Figure 9.9.2 portrays the project with those jobs crashed to minimise the project duration at a minimum premium expenditure. The same nine jobs remain critical, and to achieve completion of the project in 44 days it is necessary to crash the nine jobs indicated in the last column of the job sheet. Note that one of the critical jobs, Job 20, is not crashed; Job 22 cannot be crashed, and so there is no point in crashing Job 20.

Figure 9.10 shows the project-cost tabulation, and as the estimated rate of profit to be obtained on completion of the project is £40 per day, all jobs are crashed. Under these circumstances the project would not be crashed if the additional manufacturing capacity were not to be required before the elapse of 62 days from the commencement of work on the project. In both the charts of Figure 9.9 the time lines at the end of job lines 8 and 9 have been directed to separate the dependences of the various jobs that follow.

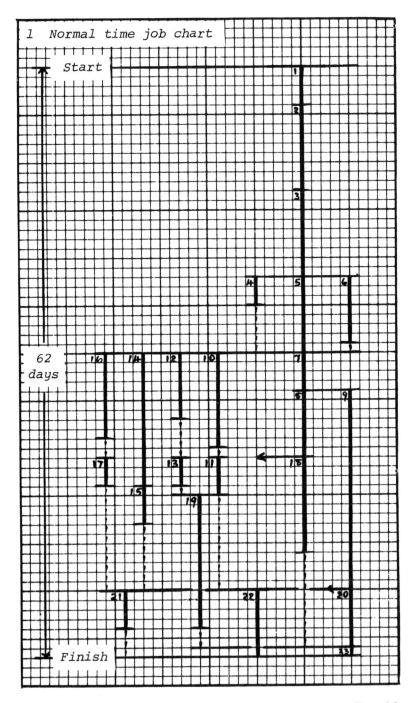

Figure 9.9

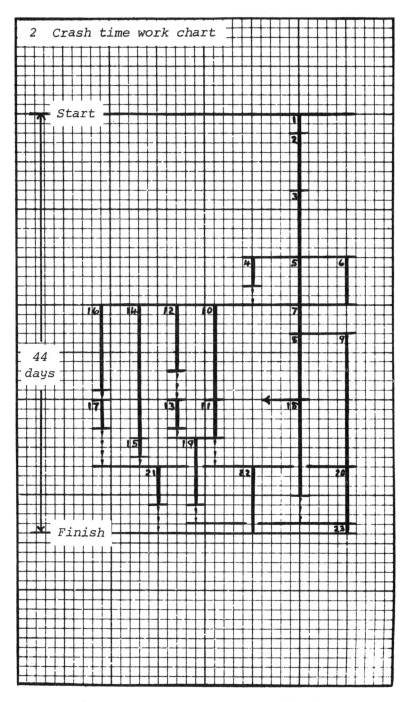

Job charts for Project 7

Jobs to be crashed	Days saved	Premium cost	Premium cost per day saved
		£	£
1	2	35	17·5
2	3	60	20
3	2	60	30
5 & 6	3	100	33·3
7	1	30	30
9, 15 & 19	7	160	23
Total	18	£445	£24.7

Figure 9.10   Project-cost tabulation for Project 7

Project 8 — Minor installation project — Installation of additional research facilities

During the course of a research investigation which it is hoped will lead to routine manufacture of a new product, it may become necessary to install pilot-plant-scale equipment, or even half large-scale equipment, to pursue the investigation and derive additional data before deciding to go ahead with full-scale manufacture. For some products and processes it will be necessary for the equipment to be specially designed and fabricated from special materials, but wherever possible any suitable equipment available in a Monsanto factory will be used. The two jobs of longest duration in such a project are likely to be those concerned with ordering and awaiting receipt of equipment and preparing a suitable building to receive the equipment — assuming that an existing building can be used. The actual installation of equipment may be a comparatively small job, as usually only a few items of equipment are involved.

Figure 9.11 gives the job sheet for a project that has just been completed. Research work has been carried out for about eighteen months, and the starting date of the project is the date when requirements for the pilot plant equipment were finalised. At this time the job sheet and job chart were prepared, and the first job of the project was allocated three weeks for preparation of the final capital appropriation request and the detailed engineering design report. This is followed by a 'job' with a time allowance of one week awaiting official approval from management to proceed with the project. The job chart for this project is constructed easily as shown in Figure 9.12, and a time scale is drawn at the side of the chart. The critical path is seen immediately to consist of Jobs 1, 2, 5, 10, 11, 13 and 18. At the time when the chart was drawn it was realised that there was very little likelihood of saving any time on Jobs 1 and 2, with only small savings possible on Jobs 10, 11, 13 and 18. The use of additional manpower or overtime working, however, could effect a considerable reduction in the duration of Job 5. But it can be seen from the job chart that the only saving in the project duration that can be made by reducing the duration of critical jobs is three weeks, after which Job 3 becomes critical. It was known that the duration estimated for Job 3 could

Job number	Job	Job sequence	Job time (weeks)
1	Prepare final draft of C.A.R.	0;1	3
2	Await approval of C.A.R.	1;2	1
3	Order and receive extruders and instruments (U.K.) and equipment from another Monsanto factory	2;3	16
4	Order and receive PMG pump	2;4	14
5	Clear site and provide new facilities for civil trades	2;5	14
6	Order and receive electrical equipment	2;6	10
7	Fabricate services pipework	2;7	2
8	Fabricate special equipment for water and PMG unit	2;8	8
9	Dismantle and modify steelwork in building 17	2;9	2
10	Modify building 17, civil work	5, 9;10	3
11	Install offices in building 17 (first part)	10;11	1
12	Install services pipework, building 17	7, 11;12	1
13	Install offices in building 17 (second part)	11;13	2
14	Complete electrical work, north end building 17	6;14	4
15	Complete electrical work, south end building 17	5, 6;15	3
16	Complete electrical work in office	11, 15;16	1
17	Install equipment	3, 4, 8, 12;17	3
18	Connect power to equipment (to start 1 week after start of 17)	14, 16, 17;18	3

Figure 9.11   Job sheet for Project 8: installation of additional research facilities

not be reduced as some of the equipment was to be imported, and it was decided, therefore, to work to standard time throughout. In the event, and despite regular progressing, delivery of some of the equipment was delayed by two weeks so the final float attaching to Job 3 was only one week. The job chart contains two interesting features. Two job end-lines are shown, each of which has a director arrow on each side of the job line. That at the end of Job line 5 is directed on each side to show that the start of Job 10 is dependent on the completion of Jobs 5 and 9, while the start of Job 15 is dependent on the completion of Jobs 5 and 6. The job end-line of Job 11 is similarly directed, to show that the start of Job 12 is dependent on the completion of Jobs 7 and 11, the start of Job 16 is dependent on the completion of Jobs 11 and 15, while the start of Job 13 follows directly after the completion of Job 11. As has been stated previously, these relationship directors are drawn on the job chart used for planning and scheduling, but are not drawn on the work chart used to control the progress of the work.

The second feature is the time line drawn after the elapse of one week of Job 17. This describes the instruction given in the job sheet whereby Job 18 may be started one week after the start of Job 17. If desired, Job 17 could be divided

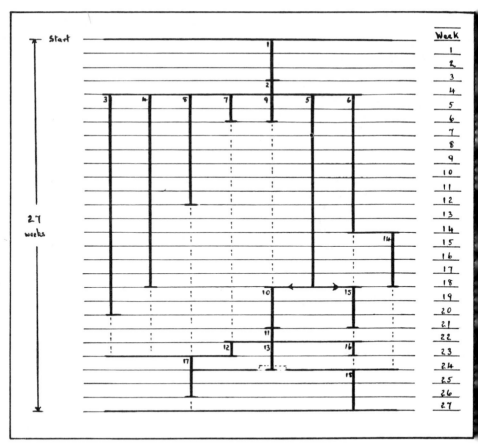

Figure 9.12 Job chart for Project 8

into two jobs, 17a and 17b, but there is no reason why a time line should not be drawn wherever the project planner or supervisor thinks fit. Very little scheduling was needed with this project, and it was made easy with the job chart. Jobs 7 and 8 were sequenced, and Jobs 14 and 15 were sequenced. The other jobs were dealt with as shown on the chart.

Project 9 — Special rehabilitation project — Centralisation of workshop facilities

Special rehabilitation projects do not occur very often, and whenever one does occur it will be the subject of much discussion among the supervisors concerned to decide exactly what work is to be done and what will be the resulting benefit. As with every other project there must be a return on capital sufficiently large to make the project worth while, and the return on capital may be derived directly from the rehabilitation project or may be derived from some other project to which the rehabilitation project is an essential preliminary.

A project such as this deals with many of the workshop facilities available to the factory, and may involve a lot of work and movement of many items of equipment on a restricted site. The main difficulty which arises in the planning stage is the determination of the sequence of jobs necessary for work to proceed, and this must be followed by determination of the sequence (or sequences) of jobs that must be achieved to ensure satisfactory completion of the project and avoid dislocation of repair and maintenance activities in the rest of the factory. This becomes quite easy when a job chart is prepared, and this project provides an example of one case where it is useful to prepare the job sheet and the job chart together. The job sheet for this project is shown in Figure 9.13, and it is seen that the project contains a variety of jobs. Two of the jobs, Jobs 8 and 14, relate to order and receipt of materials and equipment, while the other jobs are concerned with demolition, erection, modification, re-siting, installation, and decorating — a fairly reasonable

Figure 9.13 Job sheet for Project 9: centralisation of working facilities

Job number	Job	Job sequence	Job time (days)
1	Move welder foreman's office and fit temporary side wall	0;1	3
2	Demolish redundant offices in central fitting shop	0;2	5
3	Dismantle welding shop store	1;3	2
4	Resite electrical cables clear of dividing wall	1;4	12
5	Dismantle washing facilities and lavatories	29;5	13
6	Erect anti-flash curtain	2, 4;6	2
7	Demolish dividing wall between shops	5, 6;7	7
8	Procure steelwork (due 10 days from start of project)	0;8	10
9	Install welding shop extensions and modifications to welding shop building	3, 8;9	20
10	Modifications to existing electrical equipment	4, 5, 6, 7;10	5
11	Install new workshop lighting	2;11	15
12	Modify heating in workshops	4, 5, 6, 7;12	15
13	Install gantry crane	3, 8, 11;13	20
14	Procure equipment (due 20 days from start of project)	0;14	20
15	Install new offices and tool store	5, 14;15	10
16	Install new bins in tool store	15;16	3
17	Dismantle fitting shop store	16;17	6
18	Transfer blacksmith shop	17;18	4
19	Transfer apprentice training equipment	17;19	8
20	Move tradesmen's workbenches (welding shop)	0;20	2
21	Resite tradesmen's workbenches (fitting shop)	6, 20;21	2

(continued overleaf)

Figure 9.13 *(continued)*

Job number	Job	Job sequence	Job time (days)
22	Resite tradesmen's workbenches (welding shop)	15, 20;22	2
23	Install new workbenches	21;23	2
24	Install degreasing equipment	23;24	2
25	Install pump and gearbox test equipment	23;25	5
26	Install new fume exhaust system	4, 5, 6, 7;26	4
27	Resite machine saw	4, 5, 6, 7;27	2
28	Resite radial drilling machine	4, 5, 6, 7;28	5
29	Move Wilson lathe from existing site	0;29	2
30	Resite Wilson lathe	17, 29;30	2
31	Resite milling and shaping machine	19;31	2
32	Resite bending rolls	9, 13;32	2
33	Resite powered guillotine	9, 13;33	3
34	Install new plate storage rack	29, 33;34	2
35	Resite pedestal grinding machine	4, 5, 6, 7;35	2
36	Install new bar racks	15;36	2
37	Install new tube cleaning equipment	23;37	8
38	Install hand-operated trailer for steamtrap maintenance	23;38	3
39	Install oxygen header	34, 41;39	10
40	Install acetylene header	34, 41;40	10
41	Resite hand-operated guillotine	32, 33;41	1
42	Building decoration—fitters shop	1-38;42	5
43	Building decoration—welders shop	39, 40;43	5

assortment of jobs for a small project. It is assumed that 'cleaning up' is included as part of each job where cleaning up is necessary, so that there is no final job to clean up the site. The job chart for the project is drawn in Figure 9.14, and this shows that critical path to consist of Jobs 2, 11, 13, 33, 34, 39 and 40, and 43. Also note that there is another major path consisting of Jobs 14, 15, 16, 17, 19, 31 and 42 which could become critical if the delivery of equipment required in Job 14 were to be delayed by 6 days or more; this is a job for the progress clerk to look after. The other jobs all seem to have plenty of float, and it is easy to see from the job chart that many of the jobs can be sequenced to facilitate the movement of teams of men doing the actual work and simplify the job of the supervisor. For example

1 Starting on the 18th day: Jobs 21, 23, 24, 25, 35, 16, 17, 37 and 31 may be sequenced and carried out in this order, and if so, there is no float in this sequence.

2 Starting on the 22nd day: Jobs 38, 26, 27, 28, 24, 22, 33, 34 and 36 may be sequenced and carried out in this order, and if so, there is a 1-day float for the first six of these jobs.

3 Starting on the 25th day: Jobs 12, 30, 32, 41 and 10 may be sequenced and carried out in this order, and if so, there is no float in this sequence.

All three of these sequences of jobs can be completed by the end of the 49th

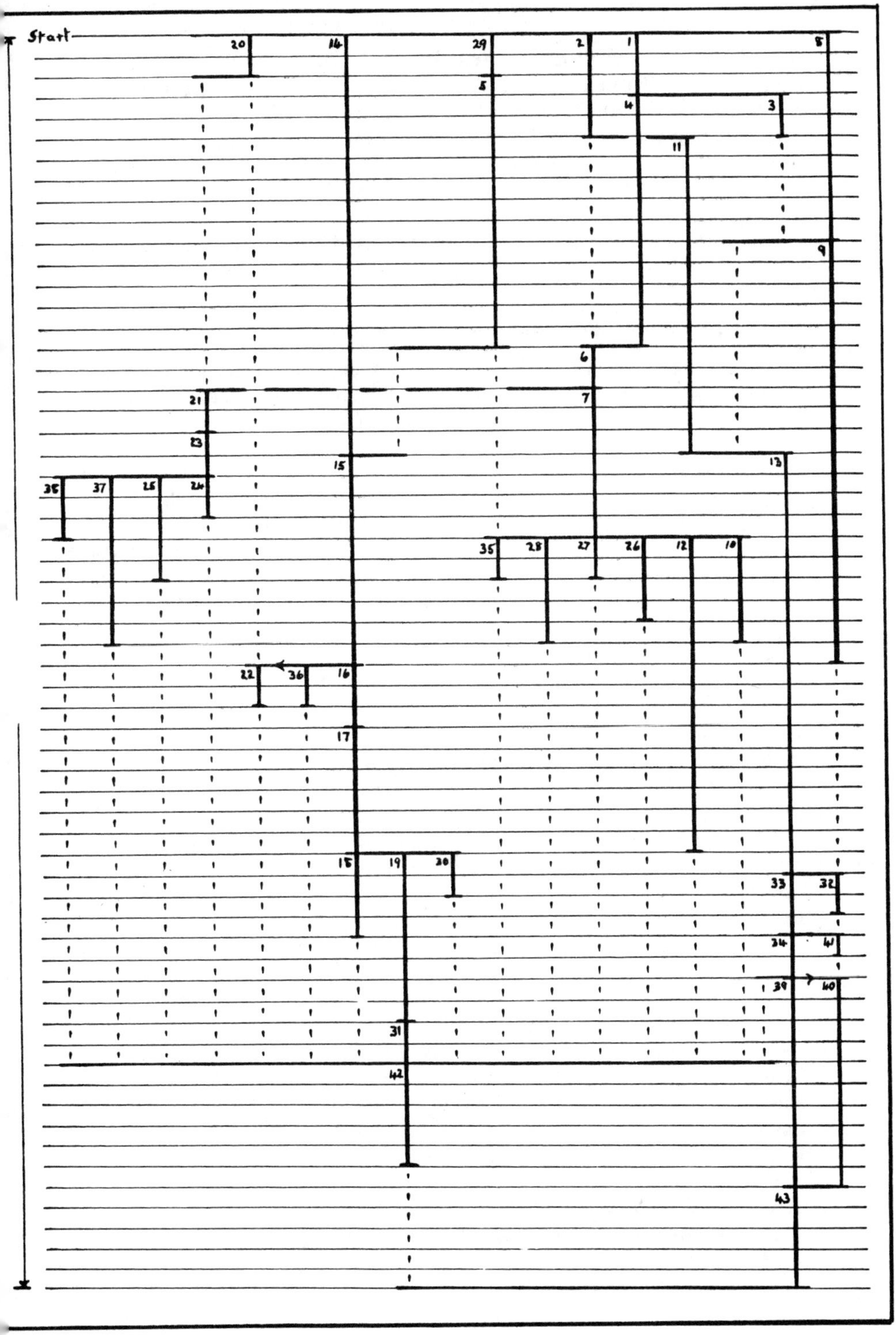

Figure 9.14 Job chart for Project 9

day, so that Job 42 can start on schedule. This ease of sequencing runs of jobs is one of the advantages of job charts that is not offered by arrow networks.

The job chart shows two directed time lines. The one at the end of job line 15 shows that each of Jobs 16 and 36 may be started when Job 15 is complete, but Job 22 can be started when both of Jobs 15 and 20 are complete. The other at the end of job line 34 indicates that both of Jobs 34 and 41 must precede Jobs 39 and 40, but only Job 34 (and not Job 41) must precede Job 42. The work chart would show the desired sequences of jobs and would not show the directed time lines.

Project 10 — Workshop project — Retubing a heat exchanger

When an item of equipment fails in service it must be removed and a replacement installed as quickly as possible so that manufacture can continue with the minimum loss. Some items, such as perforated or badly worn reaction vessels, are scrapped, but many items can be reconditioned, after which they may be used again for a further period of satisfactory service. The amount of work involved in the reconditioning may vary with the age of the item and the service to which it has been put and possibly with the future service for which the reconditioned item is proposed. For example, a 150-tube heat exchanger with one or two tubes leaking and one or two showing signs of corrosion could have a long replacement life if only these four tubes were to be replaced. But if the heat exchanger had been in heavy corrosive service for a long period and leaks had developed in a few tubes, it would be very likely that a lot of the tubes would be corroded; and for the reconditioned heat exchanger to be good enough for a further useful period in the same service, it would be advisable to fit a new tube bundle. This project is concerned with such a reconditioning. The project is carried out in the maintenance workshop, not exactly at leisure but as and when labour is available. Once the project is started, someone works on it all the time the item is on the workbench, but at times there may be several craftsmen engaged on it. A job chart is a standard part of the job instruction sheet, and this shows how the order of jobs may be manipulated so that the available manpower is used to the best advantage and is used as a progress record.

The job sheet for this project is given in Figure 9.15. The jobs are detailed and the standard job time is quoted for each job, but it must be remembered that these jobs and standard times apply only to the particular type of heat exchanger concerned in this project. Heat exchangers of different patterns will have some different job elements in their reconditioning projects, and heat exchangers of different sizes (more or less tubes, and maybe longer or shorter tubes) will require different times for jobs given the same job description. The job sheet includes one job, Job 7, which is a 'removal' job, and a nominal job time of one hour is postulated; as soon as the scrap tubes are loaded on to the transport vehicle they are of no further interest to this project, and no later job in the project is dependent on Job 7.

The job chart is drawn in Figure 9.16. The critical path is seen to consist of the fourteen jobs 10, 11, 12, 16, 17, 18, 19, 20, 21, 22, 23, 26, 27 and 28; half the number of jobs and two-thirds of the scheduled time involved in the project are attributed to the critical path. Very few of these jobs can be crashed by providing additional labour; possibly Jobs 18, 19, 20 and 21 could be crashed.

Job number	Job	Job sequence	Job time (hours)
1	Bring in H.E., hoist on to repair bench and secure	0;1	1
2	Remove water box flange bolts and remove water boxes	1;2	1
3	Clean up joint faces and tube plate	2;3	1
4	Measure and mark off cutting lines on shell	2;4	1
5	Burn through shell and tubes	4;5	4
6	Remove tubes	5;6	6
7	Remove old tubes and tube plates to scrap	6;7 x	ix
8	Clean up ends of shell and bevel to 45°	6;8	4
9	Check and clean internal baffle plates	8;9	3
10	Machine and face new tube plates	0;10	8
11	Drill tube holes in tube plate	10;11	8
12	Countersink tube holes	11;12	6
13	Cut tubes to required length	4;13	10
14	Deburr and clean tube ends	13;14	6
15	Check dimensions tube O.D. and hole diameter	14;15	3
16	Line up new tube plates and tack weld to shell	9, 12;16	4
17	Check alignment with baffle plates and weld tube plates to shell	16;17	8
18	Insert tubes and check clearances	15, 17;18	6
19	Expand tubes	18;19	8
20	Check sealing of expansion boxes and clean tube plate face and tube ends	19;20	3
21	Weld in all tubes	20;21	8
22	Hydraulic test on tubes, tube plate and shell	21;22	8
23	Clean up tube plate joint faces	22;23	4
24	Cut new tube plate joints	10;24	2
25	Drain and dry out heat exchanger	22;25	2
26	Refit end boxes	3, 23, 24, 25;26	4
27	Clean down, paint and number the heat exchanger	26;27	4
28	Remove from bench and transport to stores	27;28	1

Figure 9.15 Job sheet for Project 10: retubing a heat exchanger

Where an emergency project has to be carried out, the greatest time saving is made by completing some of the jobs, such as Jobs 10, 11, 12, 13 and 14, before the heat exchanger is removed from the plant. Certain re-arrangement of the work can be made to permit some overlap between jobs such as Jobs 5 and 6, Jobs 18 and 19, and Jobs 20 and 21.

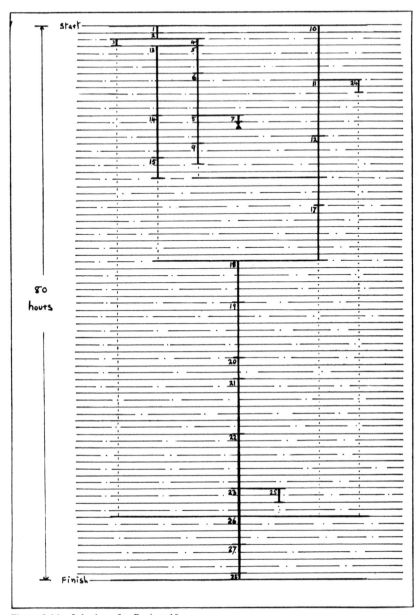

Figure 9.16 Job chart for Project 10

Chapter 10

Job Charts for Large Projects

The example project used to illustrate the techniques of Chapters 5, 6 and 7 was very small, consisting of only fourteen jobs, and the arrow network for this project was not particularly difficult to draw or to analyse. The seven examples of projects from the Monsanto factories quoted in Chapter 9 are also fairly small, and the job charts for these projects look very simple. This point cannot be emphasised too much or too often — a job chart always makes representation of a project very simple. After twenty-five years' experience in chemical plant operation, and adding the combined experience of several hundred years of our engineers, it seems that projects of this size form the great majority of the work done; and all planning, scheduling and progressing are made easy by the use of job charts.

However, on occasions, more extensive projects are undertaken. Sometimes a unit in a plant is to be rebuilt and advantage may be taken of a long shut-down to overhaul the rest of the plant. Sometimes it is convenient to shut down a plant for a week or a fortnight for 'holidays', and again the opportunity is taken to overhaul the equipment. In an overhaul project there may be sequences of jobs relating to specific units which have no job relationship with each other but which are related by manpower scheduling as a craftsman or team of craftsmen may move from a job on one unit to a job on another unit. An overhaul project may include several sequences of jobs, such as

'Examine and renew where necessary starter and motor of Reactor 1'
'Examine and renew where necessary starter and motor of Reactors 2, 3, 4, 5 and 6'

which could be carried out concurrently by six electricians — and would be if the jobs were critical — but which may be completed easily and conveniently by one electrician tackling them in succession. To simplify the job sheet and the job chart and ensure that the six jobs are brought to the attention of the persons concerned at the same time, all six jobs are considered as the compound job

'Examine and renew, where necessary, starters and motors of Reactors 1, 2, 3, 4, 5 and 6'

and the time estimate for the compound job is six times the estimate for a single inspection.

There may be many dependences in an overhaul project where one job leads directly to jobs on different units. For example, after an acid line has been drained and the necessary flanges disconnected, a series of jobs such as

17 Dismantle and renew acid supply and feed lines
18 Dismantle and reline reactor cover
19 Dismantle and overhaul acid pump
20 Dismantle and renew branch of acid headtank

can be started and four distinct sequences of jobs on four different units have developed.

In this chapter four larger projects completed in Monsanto factories will be examined, and it will be shown how job charts have been used.

Project 11 — An overhaul project

Figure 10.1 gives the job sheet for an overhaul project; the individual job names have been omitted, but many of the jobs listed were compound jobs. If the reader is interested in drawing arrow networks, this example offers a chance for practice, and the opportunity should be taken before reading any further.

The job sequence column shows that there are many multiple dependences. For instance, each of Jobs 17, 18, 19 and 20 can be started only when all Jobs 3, 11, 12 and 13 are complete. Also, there are three instances of dependences between jobs in different sequences

1   Jobs 15 and 16 may be started when Job 7 is complete, but Jobs 25 and 26 may be started only when all Jobs 7, 8, 10 and 17 are complete,
so that in an arrow network a dummy must be inserted from the event at the finish of Job 7 to the event at the start of Jobs 25 and 26.

2   Job 21 may be started when Job 13 is complete, but Jobs 17, 18, 19 and 20 may be started only when all Jobs 3, 11, 12 and 13 are complete,
so that in an arrow network a dummy must be inserted from the event at the finish of Job 13 to the event at the start of Jobs 17, 18, 19 and 20.

3   Jobs 45 and 48 may be started when Jobs 38 and 39 are complete, but Job 47 can be started only when all Jobs 36, 37, 38, 39 and 44 are complete,
and in an arrow network a dummy must be inserted from the event at the finish of Jobs 38 and 39 to the event at the start of Job 47.

In order to provide comparison between an arrow network and a job chart on a 'medium-size' project scale, the arrow network for this project is drawn in Figure 10.2. If the reader is still interested in arrow networks, calculation of the critical path and floats of non-critical jobs may provide more practice. The network with the critical path outlined is shown in Figure 13.52 at the back of the book. As in our previous simple example, derivation of informa-

Job number	Job sequence	Job time (days)	Job number	Job sequence	Job time (days)
1	0;1	4	26	7, 8, 10, 17;26	5
2	0;2	3	27	20, 21;27	3
3	0;3	6	28	20, 21;28	5
4	0;4	3	29	28;29	8
5	1;5	5	30	19, 27, 29;30	2
6	1;6	7	31	18, 26, 30;31	4
7	1;7	2	32	16, 24, 25, 31;32	3
8	1;8	8	33	22, 32;33	4
9	1;9	2	34	16, 24, 25, 31;34	2
10	2, 9;10	4	35	18, 26, 30;35	3
11	2, 9;11	7	36	18, 26, 30;36	10
12	4;12	6	37	19, 27, 29;37	7
13	4;13	8	38	19, 27, 29;38	7
14	5;14	3	39	28;39	5
15	7;15	4	40	28;40	7
16	7;16	6	41	22, 32;41	6
17	3, 11, 12, 13:17	4	42	33, 34, 35;42	4
18	3, 11, 12, 13;18	7	43	33, 34, 35;43	3
19	3, 11, 12, 13;19	8	44	33, 34, 35;44	5
20	3, 11, 12, 13;20	5	45	38, 39;45	2
21	13;21	4	46	41, 42;46	2
22	6, 14, 15:22	14	47	36, 37, 38, 39, 44;47	4
23	6, 14, 15;23	6	48	38, 39;48	8
24	23;24	4	49	40, 45;49	5
25	7, 8, 10, 17;25	6	50	49;50	3

Figure 10.1 Job sheet for Project 11: an overhaul project

tion from the arrow network is a painstaking, tedious and boring occupation, and anyone who gets this far with the planning of the project using an arrow network is well advised to use a computer for the rest of the work. However, the time and effort involved in the preparation of a first feasible plan can be reduced enormously, and planning and scheduling simplified greatly, if the network is produced in the form of a job chart. Construction of the chart proceeds in exactly the same manner as for all the previous examples, and it is not proposed to describe this construction in detail. The chart now has more lines on it than any of the charts drawn in Chapter 9, but resolution of the chart is easy.

Figure 10.3.1 shows the job chart for Project 11. The critical path may be identified without any trouble; it consists of Jobs 1, 9, 11, 20, 28, 29, 30, 31, 32, 33, 44 and 47, and extends over 53 days. The float available to individual floaters or sequences of non-critical jobs is shown clearly, and this job chart is used as the first feasible plan from which work scheduling and the final work chart are made. The duration of this overhaul extended over a period of 53 days, and it was necessary to prepare a schedule which allocated the available manpower in the best possible manner between this project and others which were carried out at the same time. Figure 10.3.2 shows the final work chart for Project 11 in which the general layout of job lines has been kept the same as in

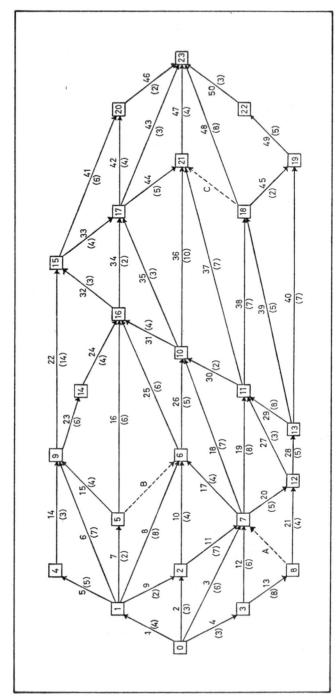

Figure 10.2 Arrow network for Project 11: A, B and C are dummies; (n) indicates job duration

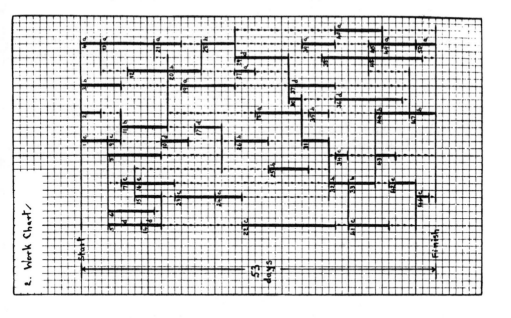

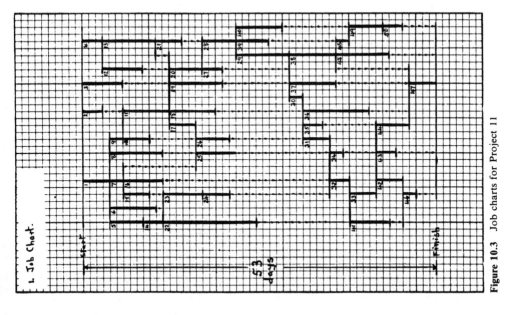

Figure 10.3 Job charts for Project 11

Figure 10.3.1 to permit easy comparison, but the positions in time of many of the jobs have been moved to show the jobs completed consecutively instead of concurrently. Jobs which fall in the same sequence by reason of work-force scheduling are lettered a, b, c or d.

It can be seen from Figure 10.3.2 that Jobs 4, 13, 21, 19, 27, 18, 39, 40, 49 and 50 are lettered 'a'. This is a sequence of jobs to be completed by a team of fitters and mates and forms a critical path through the project. Jobs 3, 11, 20, 28, 26, 25, 32, 33, 44 and 47 are lettered 'b' and lie in a sequence of jobs to be completed by a mixed team of fitters, riggers, and welders; this sequence also forms a critical path through the project. Jobs 1, 9, 7, 16, 23, 24, 22, 34, 41, 42 and 46 lie in a sequence of jobs to be completed by a second team of fitters, and this sequence has a 1-day float which can be taken in any one of three places: before Job 42, between Jobs 42 and 46, or after Job 46. A fourth team of craftsmen is required for the jobs lettered 'd', starting on the fifth day of the project and completing Jobs 5, 14 and 10, then working one day on another project, returning to Project 11 for Job 17, then moving to another project for 2 days, and finally returning to Project 11 for Jobs 29, 37 and 36. Jobs 12 and 38 were scheduled to fit in with work on other projects, and the location of Job 38 has fixed that of Job 45 and has reduced the float on Job 48 to 2 days. The positions in time of Jobs 2, 6, 8 and 43 remain unchanged, as labour could be made available for these and for the critical jobs 30 and 31 at the times indicated. Of the fifty jobs in the project only six, Jobs 2, 6, 8, 43, 46 and 48, retain float; the other jobs are all critical, twelve of them (as per first feasible plan) because of work relationships and 32 because of the necessity to schedule jobs for maximum utilisation of manpower. This scheduling was made easy by the use of job charts, and there was no difficulty in allocating manpower between this and other projects.

Project 12 — Overhaul of Plant X

A short time ago it was decided to modify the reactors in one of Monsanto's plants, and since the units were expected to be out of commission for about a fortnight, it was agreed to extend the project so that other items of equipment could be examined and overhauled. It was proposed to renew several other items which had been in service for a long time and for which early failure could be expected.

Figure 10.4 gives the job sheet for this project, and it is seen immediately that although only 47 jobs are listed, several of them are compound jobs. For example, Jobs 25 and 26 each deal with the installation of twelve items, and each could have been listed as twelve separate jobs because the items were installed individually. There are some jobs dealing with the dismantling and removal of pipework (Jobs 3-15), and replacement of the pipework or installation of new pipework (Jobs 29-43), which could have been divided into smaller job elements according to the location of the pipework. Had this subdivision been made, the project would have consisted of about 150 jobs. It is convenient for planning, scheduling and supervision to compound associated jobs as far as possible and eliminate any possibility of confusion.

In this project there are three restrictions dealing with craftsmen. First, the number of craftsmen available: the maximum allowance is 40 fitters, 8 riggers,

Figure 10.4 Job sheet for Project 12: overhaul of Plant X

Job number	Job	Job sequence	Job duration (hours)
1	Weld brackets, modify HPPS platform and remove stairway and platform above reactor 1	0;1	12
2	Remove detachable portion of south wall	0;2	4
3	Remove redundant return bends and jumpers	0;3	28
4	Dismantle line 1161	0;4	6
5	Dismantle line 1160 and remove hydrant spurs W. side	0;5	4
6	Dismantle HP line E.12 to IT.14	0;6	8
7	Remove 4″ N.B. water connection (line 22550)	0;7	4
8	Remove 2″ steam line (line 22236)	0;8	4
9	Remove 4″ water line T.1. to T.9	0;9	8
10	Remove 1″ steam line (line 22235)	0;10	4
11	Remove 1″ condensate line from E.12	0;11	8
12	Dismantle return bends on E.12	6, 8, 9, 10, 11;12	14
13	Remove 4″ water line (line 22512)	2, 8;13	8
14	Prepare for 1174 installation	1, 4, 5, 7, 13;14	8
15	Remove 1″ condensate line from T.8	0;15	8
16	Prepare reactor tubes for removal	6, 8, 9, 10, 11, 15;16	16
17	Remove 4″ water line from T.15 to PL.1120	0;17	8
18	Remove thermocouples to store	0;18	6
19	Install tee, reducer and spool piece in reactor discharge	1, 4, 5, 7, 13, 18;19	8
20	Remove RO 1 tubes	3, 5, 7, 16, 22;20	8
21	Weld new support angles to new reactor tubes	25;21	8
22	Complete fabrication of and install upper section of line 1174	14, 19;22	8
23	Resite return bends T.30/35 and T.8/13	25;23	8
24	Resite EMG bend T.14/15 and T.18/19	3, 5, 7, 16, 22;24	16
25	Install 12 HP tubes 1-8 and 31-34	20;25	12
26	Install 12 new jacketed return bends	25;26	20
27	Install 8 jacketed tubes E.12	12;27	12
28	Install E.13	20;28	20
29	Install HP pipework E.12 to Z.19	3, 5, 7, 16, 22;29	36
30	Install HP pipework 1160	27;30	12
31	Install HP pipework to E.13	20;31	28
32	Install HP pipework to IT.14	20;32	12
33	Modify platform and install spool pieces and bends	24;33	8

(continued overleaf)

Figure 10.4 *(continued)*

Job number	*Job*	*Job sequence*	*Job duration (hours)*
34	Install HP return bends and service jumpers on E.12	17, 27;34	24
35	Connect service lines 22236 and 22338 to E.12	17, 27;35	20
36	Install service lines 22233 and 22234 to E.13	25;36	16
37	Connect steam lines 22237 and 22341 to reactors	25;37	18
38	Install new steam lines 22335 and 22337 to reactors	25;38	24
39	Install new water lines 22550 and 22520 to reactors	25;39	16
40	Install new water lines 22512 and 22513 to T.19	25;40	16
41	Install 4″ water connections 1174 to 1120	25;41	8
42	Install 450 psi steam line to SBV stack	1, 4, 5, 7, 13, 18;42	24
43	Install blowdown line HPPS to VS	1, 4, 5, 7, 13, 18;43	24
44	Inspect H.P. pipework	23, 26, 28-34, 41;44	12
45	Install new thermocouples	44;45	4
46	L.P. reactor test	21, 35-40, 42-44;46	4
47	H.P. reactor test	45, 46;47	4

6 welders and 3 platers. Secondly, the number of men working in the plant at the same time should not exceed 50; if it does, the likelihood is that workmen will be getting in each other's way. Thirdly, there are some specialised jobs that are best done by the same teams of craftsmen who have acquired the greatest expertise. Figure 10.5 shows the job chart for this project; this is the first feasible plan and, of course, shows the arrangement of jobs if each is started at its earliest possible moment. Note that this plan cannot be worked because on Day 7 the labour requirement would be 48 fitters (which are not available), 8 riggers (which are available), 8 welders (which are not available), and 1 plater (which is available), making a total of 65 men in the plant, so that this plan fails on three counts.

The procedure adopted is to arrange the jobs to satisfy the third restriction given above, if possible within the 88-hour range of the critical path shown in Figure 10.5. For this:

Team '*a*' attends to Jobs 3, 29 and 40, 24 and 34.
Team '*b*' attends to Jobs 4, 18, 19, 22, 42 and 43.
Team '*c*' attends to Jobs 5 and 7, 13, 12, 27, 30 and 33 and 35, and 23.
Team '*d*' attends to Jobs 10, 9, 16, 17, 25 and 28.
Team '*e*' attends to Jobs 8, 6, 32, 41 and 38.
Team '*f*' attends to Jobs 15, 31 and 39.

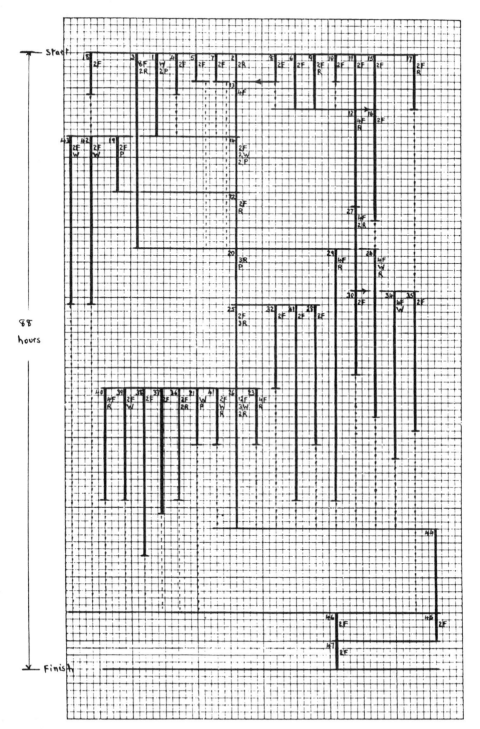

Figure 10.5 Job chart for Project 12

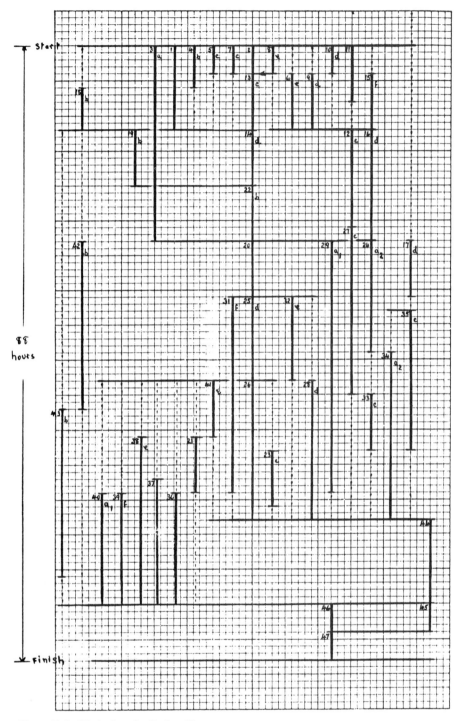

Figure 10.6 Work chart for Project 12

The other jobs are not assigned, but Jobs 36 and 37 are left as late as possible so that most of the work on Job 26 is done and the area cleared before Jobs 36 and 37 are started.

The work chart for the project is shown in Figure 10.6, and it is noted that an acceptable schedule of jobs has been obtained by allocating jobs to the work teams as listed above. All the manpower restrictions are met, and the project duration remains at 88 hours. Apart from jobs known to be critical because of their relationships, Jobs 1, 2, 3, 13, 14, 16, 19, 22, 20, 25, 26, 44, 45, 46 and 47, many of the other jobs have become critical because of the necessity to schedule manpower, and few of the other job sequences have very much float. For instance, in the series of Jobs 'c', Jobs 5, 7 and 13 are critical and the sequence formed by Jobs 12, 27, 30, 33, 35 and 23 has two hours' float which, since the team splits after Job 27, can be taken safely only with Job 23.

Project 13 — Centralisation of manufacturing facilities

From time to time instances occur where demand for a certain project increases so that the plant, A, originally built for manufacture of that product cannot cope with the demand. There may be another plant, B, which is not working at capacity and which can be used — maybe with some small modifications — to help meet the demand for the product. This arrangement might proceed for a long time, but sooner or later it becomes unsatisfactory and a decision is made to centralise manufacture of the product in its original plant, A. Because the equipment in plant B cannot be released, some other equipment has to be provided, and this may be surplus equipment from other plants or new equipment purchased for the project. Project 13 deals with one such case that has been completed recently.

The centralisation project must be carried out in three stages, and when the project is submitted for approval an estimate must be provided of the time at which each stage is likely to be complete. In the first stage, which starts as soon as the project is approved and the necessary capital allocated, all the planning work must be done. Capacity requirements must be determined: flow sheets, equipment drawings and layouts, and material requirement lists must be completed; and the necessary orders must be placed with suppliers. The second stage contains all the work that can be done before any shut-down of the plant takes place, and the third stage includes those jobs that can be done only when the plant is shut down. This third stage must be completed as quickly as possible so that manufacture can be restarted without delay. In such projects it is usual to prepare detailed job sheets and job charts to describe the work and, as happens so often, the various charts are required for different purposes. Figure 10.7 gives the job sheet for project 13, and is divided into three stages. All the jobs in the first stage, 'planning', are compound jobs, but it has been found easier to separate jobs such as Jobs 10 to 21 where the orders are to be placed on different suppliers; progressing the delivery of items may be done by different people at different times. Most of the jobs in the other two sections are compound jobs as this simplifies planning, scheduling and control of the work. The last job in the 'planning' section, Job 22, includes preparation of job charts for the rest of the project.

Figure 10.8 shows the job chart for stages 1 and 2 of the project and covers

Figure 10.7   Project 13 job sheet: centralisation of manufacturing facilities

Job number	Job	Job sequence	Job duration (weeks)
Stage 1	*Planning*		
1	Investigate capacities of melt system, refrigeration and filters	0;1	2
2	Prepare flow sheets and isometrics: approve layout	1;2	1
3	Prepare general arrangement drawing	2;3	3
4	Prepare material take-off lists (for materials ex stock)	3;4	1
5	Detail civil design and work specifications	3;5	2
6	Detail steelwork design and work specifications	3;6	1
7	Detail equipment design and work specifications	3;7	3
8	Detail pipework design and work specifications	3;8	3
9	Detail electrical and instrument design and work specifications	3;9	2
10	Order and receive 2 melt tanks	7;10	4
11	Order and receive 2 transfer pumps	7;11	5
12	Order and receive 2 autoclaves	7;12	9
13	Order and receive 2 crystallizers	7;13	8
14	Order and receive 3 filters	7;14	10
15	Order and receive 1 headtank	7;15	4
16	Order and receive 2 slurry pumps	7;16	8
17	Order and receive 3 storage tanks	7;17	12
18	Order and receive steelwork	6;18	8
19	Order and receive civil work	5;19	5
20	Order and receive electrical work	9;20	3
21	Order and receive insulation and painting	8;21	3
22	Planning; detail work orders; prepare job charts	4, 5, 6, 7, 8, 9;22	4
Stage 2	*Work prior to plant shut-down*		
23	Fabricate molten product transfer lines, other pipelines, brackets and supports	22;23	3
24	Construct pump and tank foundations	19, 22;24	1
25	Construct foundations for vessel supporting steelwork	22;25	1
26	Fabricate and erect supporting steelwork and floors	18, 22;26	2
27	Erect transfer lines; hydraulic test (last 2 days)	26;27	2
28	Steam trace transfer lines	27;28	1
29	Install 4 pumps and connect pipework	11, 16, 23, 24, 26;29	1
30	Install autoclaves, crystallizers, filters and headtank with agitators, gearboxes, etc.	12, 13, 14, 15, 26;30	2

(continued)

Job number	Job	Job sequence	Job duration (hours)
31	Install melt tanks and storage tanks	10, 17, 26 ;31	2
32	Install condensers and pipework (ex stores)	30, 31 ;32	1
33	Fabricate and install vent fan supporting steelwork	18, 23, 26 ;33	1
34	Fabricate and install ventilation ducting	22 ;34	1
35	Install vent fan and ducting	33, 34 ;35	1
36	Install electrics	20, 22, 29, 30, 35 ;36	1
37	Insulate pipelines and vessels	21, 28-32 ;37	1
38	Paint steelwork, vessels, etc.	23-27 ;38	1
39	Fabricate replacement steam main	22 ;39	2
Stage 3	Work following plant shut-down		(hours)
40	Isolate pipelines and vessels for modification	S ;40	16
41	Dismantle redundant and worn-out pipelines	40 ;41	20
42	Install molten product transfer lines (old and new plant)	41 ;42	20
43	Cross-connect circulation lines (old and new plant)	41 ;43	24
44	Modify existing vessels and fit new coils	41 ;44	40
45	Remove existing headtank (CT-74)	41 ;45	16
46	Transport CT-74 to workshop and rig in place	45 ;46	4
47	Install new supporting steelwork and floors	45 ;47	8
48	Install new head tank (CT-484)	47 ;48	12
49	Modify CT-74 and fit new coil	46 ;49	24
50	Install modified CT-74 on new foundations	47, 49 ;50	4
51	Connect new pipework to both headtanks	48, 50 ;51	20
52	Erect new steam main and connect to vessels, tracers, etc.	42, 43, 44, 51 ;52	48
53	Dismantle centrifuges, chutes and pipework	S ;53	12
54	Dismantle centrifuge supporting steelwork	53 ;54	16
55	Erect new centrifuge steelwork (some old steelwork)	54 ;55	8
56	Install centrifuges in new location	55 ;56	20
57	Install pipework, feed chutes and discharge chutes	47, 56 ;57	40
58	Install new electrical equipment and supports	51, 57 ;58	40
59	Install new instrumentation, panels and supports	52, 58 ;59	32
60	Insulate vessels and pipelines (in old plant)	52 ;60	24
61	Paint all equipment (old plant)	59, 60 ;61	16
62	Clean up site	61 ;62	8

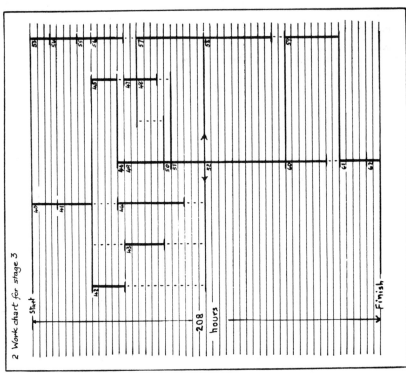

2 Work chart for stage 3

1 Job chart for planning and preliminary work

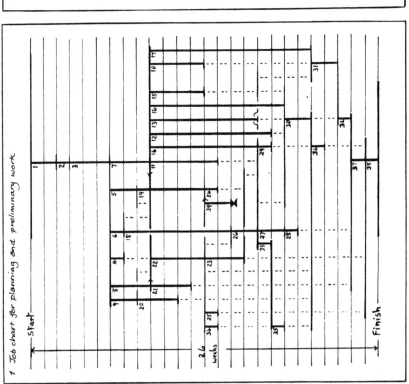

110

Figure 10.8 Job charts for Project 13

all the planning and site-work prior to the plant shut-down; all this work must be completed before the plant is shut-down, and it is seen from the chart that this part of the project will require 26 weeks. The critical path consists of Jobs 1, 2, 3, 7, 17, 31, 32, 37 and 38; and of these Job 17, for order and delivery of three new storage tanks, requires 12 weeks. Figure 10.8 includes one job, Job 39, which is not related to any other later job on this chart, but which is an essential preliminary to stage 3 of the project. The actual date agreed for shut-down of the plant may bear little relationship to the date of completion of stage 2 of the project, except that it cannot be earlier. It may be controlled solely by the time taken by both plants A and B to build up sufficient inventory to cover the shut-down period. Thus the starting point for stage 3 in the job sheet is labelled 'S' — the shut-down point — and it is understood that all jobs in stages 1 and 2 must be completed at S.

Figure 10.8.2 shows the work chart for stage 3 of the project. Note that whereas the unit of duration for the jobs in stages 1 and 2 is one week, in stage 3 the job durations are quoted in hours; and it is convenient in the work chart for one line space to represent 4 hours. The work chart shows some jobs sequenced because of manpower availability; for example, one team attends to Jobs 40, 41, 42, 43, 50, 51 and 52. Also, it is convenient not to start Job 44 until Job 45 is finished. The director arrows on either side of job line 51 indicate that Job 51 is related to job sequences shown on each side of job line 51. The critical path is seen to consist of Jobs 40, 41, 45, 46, 49, 50, 51, 52, 59, 61 and 62 and extends over 208 hours. Because one team of craftsmen is required to attend to Jobs 46, 47 and 48, an 8-hour float is introduced between Jobs 56 and 57, and Job 57 becomes critical. Thus of the 23 jobs in stage 3 of this project, twelve are critical, and of the others only two, Jobs 43 and 44, have long floats.

Project 14 — A major construction project

Up to this point we have considered projects which must be completed in a comparatively short period of time and on which, because manufacture is reduced or maybe stopped, close supervision and control are exercised to ensure that the projects are completed to schedule. These projects have been planned, scheduled, and controlled by means of job charts and, wherever necessary, the work charts derived from resource allocation (which has required manpower arrays in most cases). If any corrections or amendments are found to be necessary, the charts can be altered very easily to match the new conditions. Any more sophisticated planning technique would be a waste of time and effort if indulged in for this type of project, and would not improve planning, scheduling, or control of the work, or increase understanding of critical path principles.

However, from time to time very large projects arise and these too must be planned, scheduled, and controlled. These projects may involve several thousand jobs, which may be spread over a long period of time and which may involve many workmen on widely separated locations. There may be a large number of relation-links between jobs which tend to complicate any network representation of the project. But it remains true that if the only requirement of the network is to determine the critical path and show the time-

Figure 10.9 Job sheet for Project 14: a major construction project

Job number	Job sequence	Job time (weeks)	Job number	Job sequence	Job time (weeks)
1	0;1	2	42	36;42	4
2	1;2	2	43	28, 40;43	5
3	2;3	3	44	42;44	3
4	2;4	3	45	38;45	10
5	3;5	2	46	37;46	13
6	4;6	4	47	30, 32, 33, 44;47	5
7	5;7	5	48	43;48	4
8	5;8	3	49	41;49	8
9	6;9	4	50	41;50	12
10	8;10	6	51	39, 46, 47, 48, 49;51	5
11	5;11	5	52	39, 46, 47, 48, 49;52	4
12	5;12	4	53	[529];53	2
13	5;13	3	54	52, 53;54	3
14	5;14	5	55	50, 52, 53;55	4
15	14;15	7	56	4;56	3
16	14;16	4	57	56;57	10
17	9, 13;17	4	58	56;58	4
18	12;18	2	59	57;59	9
19	12;19	8	60	58;60	3
20	7;20	4	61	58;61	7
20a	20;20a	4	62	58;62	3
21	18;21	2	63	58;63	4
22	18;22	4	64	60;64	4
23	18;23	7	65	64;65	3
24	4;24	8	66	26, 65;66	6
25	24;25	5	67	62;67	5
26	10, 20a, 25;26	4	68	63;68	3
27	10, 20a, 25;27	4	69	63;69	4
27a	27;27a	4	70	26, 61, 65;70	1
28	19;28	9	71	26, 61, 65;71	2
29	17, 21;29	8	72	71;72	3
30	15;30	11	73	68;73	3
31	27a, 29;31	2	74	67;74	1
32	31;32	2	75	70;75	3
33	22;33	5	76	74;76	5
34	23;34	5	77	69;77	3
35	34;35	5	78	69;78	5
36	31;36	6	79	72;79	4
37	16;37	11	80	73;80	5
38	16;38	12	81	76;81	3
39	32, 33;39	13	82	75;82	3
40	32, 33;40	3	83	78;83	4
41	32, 33;41	6	84	78;84	6

(continued)

Job number	Job sequence	Job time (weeks)	Job number	Job sequence	Job time (weeks)
85	59, 79, 77, 82;85	4	128	121, 124, 125;128	3
86	77, 82;86	5	129	115;129	5
87	86;87	4	130	115;130	4
88	80, 81, 86;88	9	131	128;131	5
89	77, 82, 83;89	4	132	129, 131;132	4
90	77, 82, 83;90	6	133	130;133	5
91	85;91	3	134	127, 128;134	3
92	89;92	7	135	132;135	5
93	89;93	4	136	133;136	5
94	90;94	6	137	134;137	11
95	93, 94;95	2	138	136;138	5
96	91, 92;96	4	139	4;139	4
97	88, 90, 95, 96;97	4	140	139;140	4
98	88, 90, 95, 96;98	4	141	140;141	5
99	93, 94;99	5	142	140;142	8
100	97;100	3	143	140;143	7
101	99, 100;101	2	144	142;144	6
102	84;102	21	145	144;145	2
103	101, 102;103	3	146	142;146	7
104	4;104	3	147	141;147	15
105	104;105	4	148	141;148	7
106	105;106	7	149	145;149	4
107	105;107	6	150	149;150	5
108	105;108	5	151	146;151	8
109	105;109	5	152	143;152	8
110	105;110	5	153	150;153	3
111	26, 106;111	5	154	147, 153;154	7
112	107;112	5	155	87, 148, 151;155	6
113	108;113	4	156	152;156	15
114	108;114	4	157	155;157	4
115	109;115	16	158	157;158	3
116	112;116	3	159	152;159	4
117	113;117	11	160	157;160	5
118	114;118	4	161	154, 158;161	5
119	110;119	6	162	155, 156, 160;162	3
120	110, 118;120	3	163	159;163	4
121	111;121	4	164	150, 157, 162;164	2
122	111;122	2	165	163;165	3
123	116, 120;123	3	166	161;166	3
124	122, 123;124	4	167	166;167	4
125	116, 119, 120;125	5	168	164;168	3
126	117;126	5	169	165;169	6
127	121, 124, 125, 126;127	5	170	168, 169;170	3

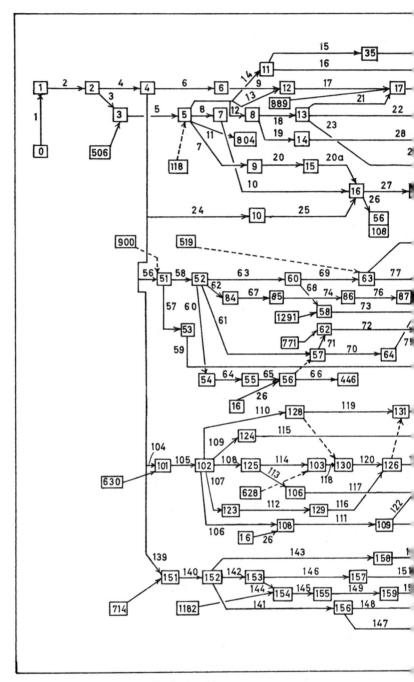

Figure 10.10 Arrow network for Project 14

114

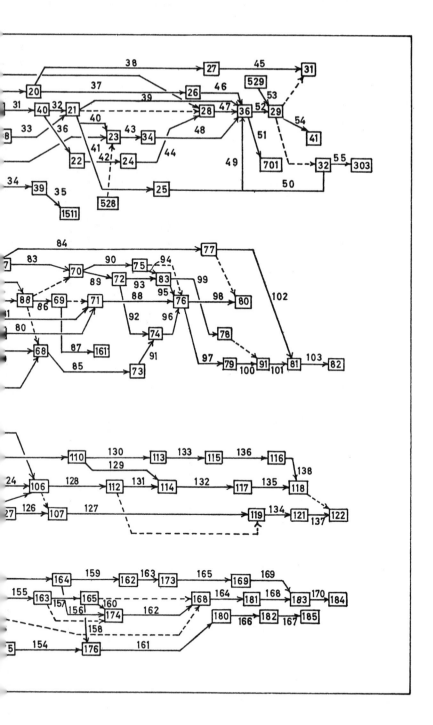

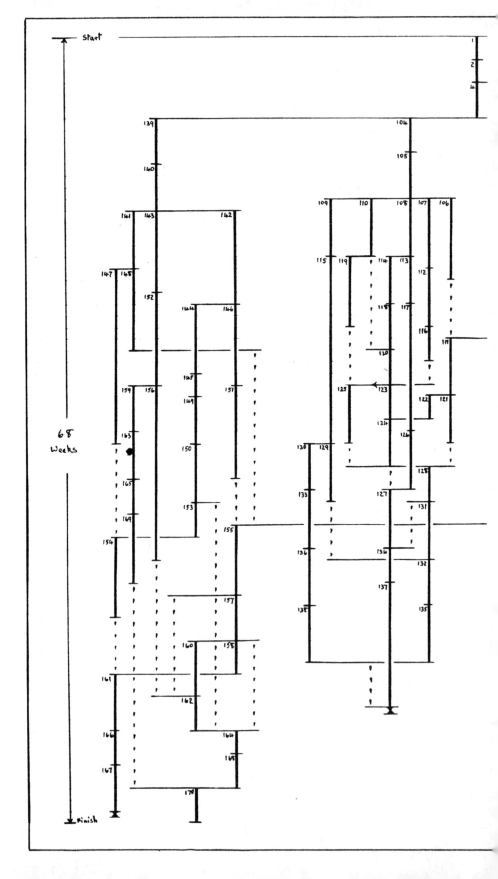

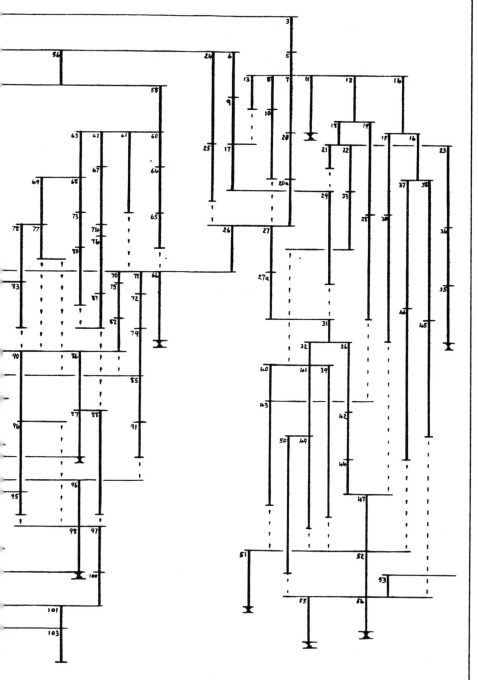

Figure 10.11 Job chart for Project 14

relationships between jobs, as these exist at the start of the project, then a job chart of whatever size may be deemed necessary is the easiest and quickest method of doing this.

Figure 10.10 shows the arrow network for a small part of one large project recently completed by Monsanto Chemicals. This has been drawn on a small scale to suit the format of this book, and to avoid undue congestion only the job numbers have been added to the network and dummies have not been numbered. Full details of the job sequences and job durations are given in Figure 10.9. The arrow network shows all the sequence relationships between the jobs in this subsection of the project and also indicates linkages to jobs in other subsections (such as the lead-in from event $\boxed{889}$ to event $\boxed{17}$ and the lead-out from event $\boxed{39}$ to event $\boxed{1511}$), but provides no other information. It is very often the case, as it is with this project, that parts of the arrow network fall into distinct compartments with some links between the compartments, and for this particular subsection four separate networks could have been drawn. In the compilation of Figure 10.9 all leads-in from other subsections of the project (except that of Job 53 from event $\boxed{529}$) and all leads-out to other subsections of the project are omitted. Since this job sheet is required to describe the construction of the job chart, dummies also have been omitted.

The job chart for this subsection of the project is shown in Figure 10.11. A job chart is intended to provide all the information that exists about a project, so that a somewhat more generous width has been used; this is not an apology but a statement of deliberate intent. But whatever scale is used, it remains a fact that the project is much better portrayed by the job chart than by the arrow network, and no difficulty is encountered in drawing a job chart, however large the project may be. Figure 10.11 is constructed in the usual manner, and some of the time lines are directed. Where the arrow network indicates a lead-out to another subsection of the project, the relevant job line in Figure 10.11 is closed by a gate. It is likely that if a job chart were to be drawn for the whole project, the leads-in to the subsection shown in Figure 10.11 might cause an extension of the (apparent) project duration. For example, in the set of Jobs 104-138, nine of the jobs are critical, but Job 124 is not one of these. If the job which ends in event $\boxed{367}$ finishes later than the starting time of Job 124 shown on the chart, then Job 124 and all the jobs dependent on it could become critical.

However, it is necessary to decide exactly what is required from a graphical representation of the project. As stated above, if all that is required is to determine the list of critical jobs and the relationships between jobs as estimated at the beginning of the project, then a job chart provides the information with complete satisfaction and no further processing of the data is required. But with a project expected to extend over 68 weeks — or maybe longer, depending on the other subsections of the project — it may be desirable, or even essential, to provide a weekly presentation of the status of the project. This would show which jobs have been finished to time, which have been delayed and by how much, which jobs remain critical, and which jobs, if any, have become critical because of delays. Variations from the original estimates in job starting and finishing times may need re-assessment of resource requirements, and it is not feasible to redraw a very large job chart and make the necessary re-assessments every week unless a very large staff is available to do the work. It is grossly uneconomic to obtain a large staff to do this work, and this is the sort of work for which a computer must be used.

The initial description of the project should be made by a job sheet and a job chart, as these permit discussions of the work by all the people concerned. Then the initial data, and any corrections as and when these become known, must be fed to a computer which will provide the updated status of the project in a very short time. Standard programs to do this are available for all the modern computers; and the more sophisticated programs accept restrictions imposed by the availability of manpower and other resources, and provide new schedules in the form of job charts (more accurately, work charts), as well as providing up-to-date information regarding the costs incurred in the project. The author believes that this is the real application for computers in the field of critical path analysis.

Chapter 11

Job Charts for Other Industries

In the previous chapters the applications described for job charts have been derived only from the chemical industry, and the examples quoted to demonstrate the uses of job charts are all from Monsanto Chemicals. But a job chart can be used to describe any project which consists of a number of distinct jobs, each bearing some time-relationship with the others, regardless of the number of jobs in the project, or the number of workmen to be employed on the project, or of any restraints such as may be imposed by resource availability. If all the jobs in a project are to be completed by one man, the project duration is the total of the durations of the individual jobs and the usefulness of a job chart is that it helps to determine the best order of doing the job. This chapter will show how the job charts can be applied to an assortment of projects in different industries, and in the customary manner a very simple domestic example will be demonstrated first, then proceeding to more extensive examples. Whatever the project may be and whatever jobs may be involved, the construction of the relevant job chart remains the same.

Project 15 — Redecorating the sitting-room

This is the sort of project that hits most people at some time or another, and whereas the job content of similar projects may differ widely according to the size of the room and one's personal inclinations for the materials to be used, there is not a lot of difference in the kind of work which has to be done. The remarks made by wives, too, are very similar; they are always unable to understand why the job took so long. Maybe a sympathetic approach with a job chart would increase their understanding and show them how the project duration could have been shortened.

When starting to consider the project, the first task should be to draw up a job sheet, and for this example note the job sheet of Figure 11.1. I have been

Job number	Job	Job sequence	Job time (hours)
1	Take out furniture and carpets, cover any left in; cover floors	0;1	2
2	Wash ceiling	1;2	2
3	Fill in cracks in ceiling	2;3	$1\frac{1}{2}$
4	One coat of paint on ceiling	3;4	$1\frac{3}{4}$
5	Clean down woodwork	1;5	1
6	Undercoat on woodwork	5, 8;6	$2\frac{1}{4}$
7	Top coat on woodwork	6;7	$2\frac{1}{2}$
8	Strip old wallpaper	1;8	$2\frac{3}{4}$
9	Rub down walls and fill in cracks	8;9	$1\frac{1}{2}$
10	Size walls	9;10	$\frac{1}{2}$
11	Hang new wallpaper	4, 7, 10;11	4
12	Clean up	11;12	$\frac{1}{2}$
13	Replace furniture and carpets	12;13	1

Figure 11.1 Job sheet for Project 15: redecorating the sitting-room

assured that this is a feasible job sheet. My own preference would be for Jobs 5 and 8 to start after the completion of Job 2 instead of Job 1, since I feel that if I were doing Job 2 it would be dangerous for anyone else to be in the room at the same time. However, accepting Figure 11.1 as the agreed sequence of jobs, it is obvious that if all the work is to be completed by one person, the project duration will be $23\frac{1}{4}$ hours, and it is necessary to examine the project a little more carefully to see if any savings in time can be made. Quite obviously, savings in time can be made only if at least one other person can be persuaded to do some of the work, and the job chart show how best this can be done. The chart is shown in Figure 11.2.

The job chart shows the critical path to consist of Jobs 1, 8, 6, 7, 11, 12 and 13 which, if the project could be completed as described by the chart, would mean that it could be finished in 15 hours. This would reduce the project by the $8\frac{1}{4}$ hours attributable to removal of Jobs 2, 3, 4, 5, 9 and 10 from the single-path project, and it is here that subterfuge and subtlety enter the scene. If Job 1 can be completed in 1 to $1\frac{1}{2}$ hours, and Job 5 can be scamped so that both these jobs are completed in 2 hours, then the project can be resolved into a dual-path project; and one needs only to persuade or press one's wife into service to achieve the 15-hour project. The wife is required to attend to Jobs 8, 6 and 7 (keeping a tight schedule), while oneself attends to Jobs 2, 3, 4, 9 and 10 (taking full advantage of the residual quarter of an hour float) before continuing on Jobs 11, 12 and 13. If the wife can help out on Job 11 so much the better, and the project duration may be reduced below the 15-hour level.

Of course, if a third person can be brought into the project the work schedule can be simplified still further. On completion of Job 1 one person can attend to Jobs 5, 2, 3 and 4 in that order; the second person can attend to Jobs 8, 6 and 7 in that order; and the third person can come in and attend to Jobs 9 and 10 as and when he has the time to spare as these two jobs have $2\frac{3}{4}$ hours' float. Then all three people can tackle the remaining jobs.

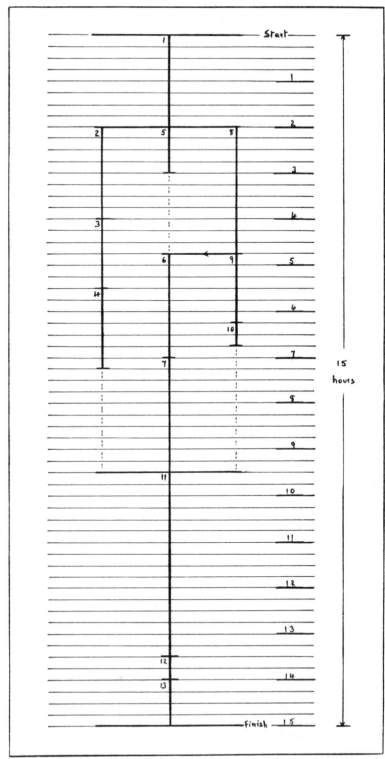

Figure 11.2 Job chart for Project 15

This technique may be extended to many domestic duties, such as changing the wheel on a car. But care is needed to prevent one's spouse from demonstrating how many pairs of hands she needs to prepare breakfast!

Project 16 — Single item manufacture: manufacture of an ornamental wooden door

There are still many items manufactured by various firms in this country which are custom-built, i.e. an item is made to special requirements detailed in a specific order. Some of these items, such as the ornamental door of Project 16, may show considerable variations between successive orders, while others, such as the mild steel vessel of Project 17, may show only minor variations between successive orders. But in all cases each order has to be dealt with independently. The job sheet for this project is given in Figure 11.3, and job times are included for both normal and crash working. The need for crash working arises when an order is accepted for urgent delivery; or when orders begin to accumulate; or when craftsmen are available, temporarily, between other jobs. Crashing can be effected only by putting more craftsmen on the project. It is convenient to have each door made by two teams of craftsmen, the 'team' consisting of one or two craftsmen (plus an apprentice sometimes) as desired.

Figure 11.4 shows the work chart for normal time working. Job 2 is to be completed in the last two hours occupied by Job 1, so that the critical path through this project consists of Jobs 1 and 2, 4, 6, 8, 9, 10, 13, 14, 15, 16, 17, 23, 24 and 25; and the project duration is seen immediately to be 62 hours. Fifteen of the twenty-five jobs in the project are critical; Job 1 is an office job, after which one team attends to the critical jobs listed above, while the second team attends to the sequence of non-critical jobs. The work chart shows that the non-critical sequence has a total float of 11 hours, which can be split into a maximum of 4 hours anywhere in the sequence of Jobs 3, 5, 7, 11 and 12, and 7-11 hours anywhere in the sequence of Jobs 18, 19, 20, 21 and 22. The work chart shows the ten non-critical jobs to be completed in succession, with no float between any of the jobs, so that the second team can be withdrawn from the project and transferred to another project at the earliest possible moment. In this case, of course, every job in the project is critical.

Figure 11.4.2 shows the work chart for crash working, and the project duration is seen to be 43 hours. The main critical path consists of the same jobs as in Figure 11.4.1, but with the restriction that only two teams of men be employed on the project, Jobs 3, 5, 7, 11 and 12 are now critical. By crashing Job 18, the sequence of Jobs 18, 19, 20, 21 and 22 is seen to have 3 hours' float, so that Job 18 can be accommodated at its normal duration, still leaving a float of 1 hour for this sequence. The decision whether or not to crash Job 18 will depend on how soon the team is required for another project. The job chart for this project showed clearly how each job must be scheduled to agree with the manpower restrictions and to complete the project in minimum time.

Job number	Job	Job sequence	Job Time normal (hours)	crash (hours)
1	Prepare design and detail drawings	0;1	10	6
2	Obtain materials from store (last 2 hr. of Job 1)	0;2	2	2
3	Rough machine posts and lintel	1, 2;3	4	4
4	Cut to size frame members	1, 2;4	6	3
5	Cut to size other members	1, 2;5	4	2
6	Plane frame members to finished size	4;6	6	3
7	Plane other members to finished sizes	5;7	4	4
8	Cut joints in frame members	6;8	4	4
9	Cut all openings, etc. for door furniture	8;9	4	2
10	Assemble frame and secure	9;10	4	4
11	Cut glass lights and infill	1, 2;11	3	2
12	Machine all beadings	1, 2;12	6	4
13	Fit all glass, infill, beadings and other members	7, 10, 11, 12;13	8	4
14	Check all door measurements	13;14	2	2
15	Finish sanding all over door	14;15	6	4
16	Paint with sealer and under-coat	15;16	4	4
17	Fit all door locks, etc.	16;17	3	2
18	Machine door posts to size	3;18	6	4
19	Cut holes and recesses for furniture	18;19	3	3
20	Check all sizes and measurements	19;20	1	1
21	Finish sanding all over posts	20;21	2	2
22	Paint with sealer and under-coat	21;22	3	3
23	Assemble door and posts and check clearances	17, 22;23	2	2
24	Fit hinges and check clearances	23;24	2	2
25	Transport from shop to store to await delivery	24;25	1	1

Figure 11.3 Job sheet for Project 16: manufacture of an ornamental wooden door

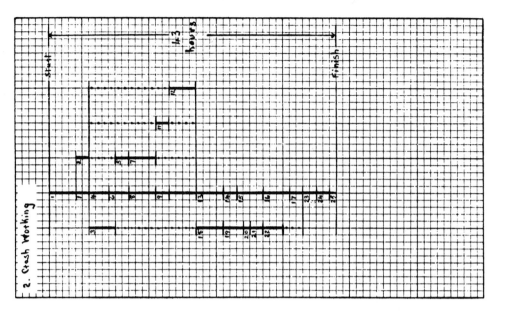

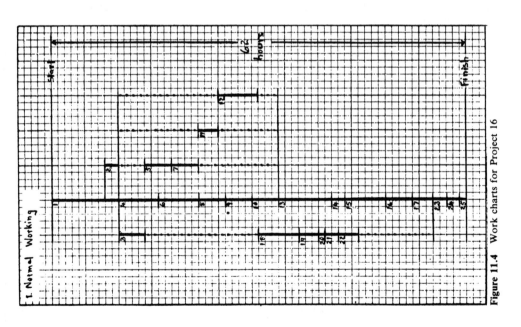

Figure 11.4 Work charts for Project 16

Project 17 — Single item manufacture: manufacture of a mild steel vessel

The job sheet for this project is given in Figure 11.5, and here again job times for both normal and crash working are included. The need for crash working in this case arises most often through unexpected failure of a vessel at a customer's works, though the reasons quoted under project 16 are applicable sometimes. The crash working time for a job is controlled by the number of men who can be physically accommodated on the job.

Some of the jobs listed in the job sheet are compound jobs, and could be subdivided at any time if this is thought advisable. For example, each of Jobs 13 and 14 concerns two flanges, so that each could be split into two concurrent jobs, which presumably would halve the quoted job times. Jobs 21, 22, 23, 24 and 26 similarly could be subdivided, but whereas the subdivisions of Jobs 21, 22, 23 and 24 would run concurrently, the subdivisions of Job 26 would run consecutively. Dividing Job 23 into two 2-hour jobs would reduce the critical path duration by 2 hours. The 'examination' parts of Jobs 15, 18, 19 and 20 can be split off from the rest of the job, but in each case would have to run consecutively to the performance of the task. Figure 11.6.1 shows the job chart for normal working. The critical path is seen to consist of Jobs 1, 3, 5, 8, 11, 15, 18, 19, 23, 25, 26, 27 and 28, and these thirteen jobs control the overall project duration at 62 hours. There is not very much float available to either of the sequences Jobs 6, 9, 12 and 16 or Jobs 4, 7, 10, 17, 20 and 24; but obviously Jobs 13, 14, 21 and 22 can be sequenced, in that order, if so desired.

Figure 11.6.2 shows the job chart for crash working in which jobs have been crashed as required to provide the minimum time plan. All jobs except Jobs 2, 13, 14, 21 and 22 are now critical and the project duration has been reduced to 48 hours. In order to sequence Jobs 13, 14, 21 and 22 on this plan, it would be necessary to crash Job 21 (which gives a saving of 8 hours) or to subdivide whichever of these jobs can be subdivided most conveniently — which may resolve into crashing Job 21. Use of the job chart ensures that the right decision is made regarding which jobs should be crashed and which jobs may be sequenced.

Building projects

Building projects vary from very small ones, where one man builds himself a shed, to very large ones, where many men of a variety of trades or crafts may be employed on a massive project on one site or on lesser projects on different sites and under several different supervisors. So that the work may proceed efficiently it is important to know exactly when and where craftsmen and equipment are required for each particular job or sequence of jobs, so that no craftsmen are idle as a result of faulty planning, and delays in work performance are kept to an irreducible minimum.

Job charts and the work charts derived from them can be of assistance to supervisors in the building industry as they

1   Facilitate planning, scheduling and control of site operations.
2   Permit overall control of operations on several sites and facilitate transfers between sites.

Figure 11.5 Job sheet for Project 17: manufacture of a mild steel vessel

Job number	Job	Job sequence	Job Time normal (hours)	crash (hours)
1	Check design drawings against order	0;1	4	4
2	Check material availability	0;2	2	2
3	Transport material to workshop	1, 2;3	2	2
4	Set up plates for marking (top dish)	3;4	2	2
5	Set up plates for marking (side wall)	3;5	1	1
6	Set up plates for marking (bottom dish)	3;6	1	1
7	Mark out plate from drawing (top dish)	4;7	9	5
8	Mark out plates from drawing (side wall)	5;8	6	4
9	Mark out plates from drawing (bottom dish)	6;9	6	4
10	Burn out plates as marked (top dish)	7;10	7	4
11	Burn out plates as marked (side wall)	8;11	4	4
12	Burn out plates as marked (bottom dish)	9;12	4	2
13	Roll flange rings	1, 2;13	6	4
14	Machine and drill flanges	13;14	10	8
15	Weld platework, side wall; examine welds	11;15	8	4
16	Form bottom dish	12;16	6	6
17	Form top dish	10;17	6	6
18	Weld bottom to side wall; examine welds	15, 16;18	12	8
19	Fit and weld flange to side wall; examine welds	14, 18;19	8	6
20	Fit and weld flange to top dish; examine welds	14, 17;20	8	6
21	Fabricate flanged branches	1, 2;21	16	8
22	Cut joints for all flanges	1, 2;22	4	2
23	Weld branches side and bottom; fit blanks	19, 21;23	4	2
24	Weld branches to top dish; fit blanks	20, 21;24	8	4
25	Assemble vessel and bolt up	22, 23, 24;25	4	2
26	Hydraulic test, dry and clean vessel	25;26	6	6
27	Paint and number vessel	26;27	2	2
28	Transfer vessel to store	27;28	1	1

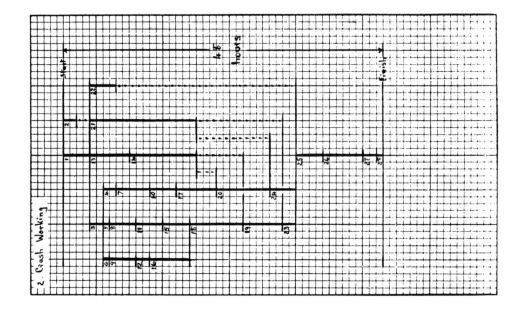

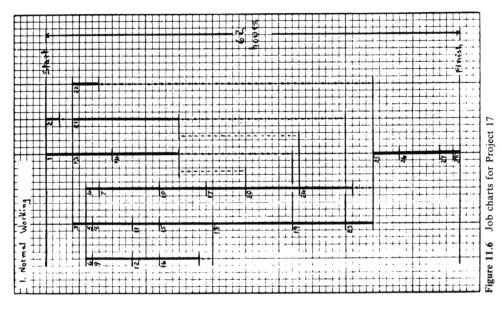

Figure 11.6 Job charts for Project 17

3   Permit assessments to be made during the design stage of the require-
    ments resulting from different designs so that the best design can be
    chosen for submission.
In all cases the charts enable planners and supervisors to see through a
project; they can see what should be done and why it should be done.
    Job charts have been drawn for some small building projects. But it must be
emphasised again that there is no difficulty in drawing charts for large
projects — larger sheets of paper will be required. Also, the time scale used in
the charts can be varied according to the use for which the chart is required.
For overall control a unit of a week might be satisfactory, whereas the same
chart used for day-to-day site control would be drawn with a unit of one day
or even a half-day, and this chart would be used as a performance record. In
the job charts which follow, director arrows have been omitted from time
lines as these only indicate job relationships and are not necessary to show the
work flow, manpower requirements, or the allocation of any other resources.

Project 18 — Construction of a concrete bungalow

Figure 11.7 gives the job sheet for this project, and the first thing to be noticed
is that most of the jobs have been divided into sub-jobs or sections to show
that several different jobs may be pursued concurrently. For example, after
clearing the site the next job to be done is to pile the ground. This can be
treated as one single job which is to be completed before the pile capping is
started, and so on, but working in this manner will give a maximum duration
project. So as soon as a sufficient area of the site is piled, the pile capping can

Figure 11.7   Job sheet for Project 18: construction of a concrete bungalow

Job number	Job	Job sequence	Job time (weeks)
1	Clear site	0;1	½
2	Pile first section	1;2	1
3	Pile second section	2;3	1
4	Pile caps, first section	2;4	2
5	Pile third section	3;5	1
6	Pile caps, second section	3, 4;6	1
7	Ground beams, first section	4;7	2
8	Pile caps, third section	5, 6;8	1
9	Ground beams, second section	6, 7;9	2
10	Filling, first section	7;10	½
11	Ground beams, third section	8, 9;11	1
12	Filling, second section	10;12	½
13	Blinding, first section	10;13	½
14	Filling, third section	11, 12;14	½
15	Blinding, second section	13;15	1
16	Base slab, first section	13;16	1½
17	Blinding, third section	14, 15;17	½
18	Base slab, second section	16;18	1

(continued overleaf)

129

Figure 11.7 *(continued)*

Job number	*Job*	*Job sequence*	*Job time (weeks)*
19	Columns, first section	16;19	1
20	Base slab, third section	17, 18;20	1½
21	Columns, second section	19;21	1
22	External work (drains, drives)	20, 21;22	7
23	Window heads, first section	19;23	2
24	Spine beams, first section	19;24	1
25	Columns, third section	20, 21;25	1
26	Window heads, second section	23;26	1½
27	Spine beams, second section	24;27	1
28	Purlins, first section	23, 24;28	½
29	Window heads, third section	25, 26;29	2½
30	Spine beams, third section	25, 27;30	2
31	Purlins, second section	26, 27, 28;31	½
32	Install heating equipment	28;32	7
33	Exterior infill brickwork, first half	28;33	½
34	Woodwool, first section	28;34	½
35	Woodwool, second section	31, 34;35	½
36	Purlins, third section	29, 30, 31;36	½
37	Roof slab, first section	31, 34;37	1½
38	Woodwool, third section	35, 36;38	½
39	Roof slab, second section	37;39	1
40	Roof screed, first section	37;40	½
41	Roof slab, third section	38, 39;41	1
42	Roof screed, second section	40;42	½
43	Roof asphalt, first section	40;43	½
44	Roof screed, third section	41, 42, 44	½
45	Roof asphalt, second section	43;45	½
46	Load-bearing brickwork, first part	41;46	½
47	Load-bearing brickwork, second part	46;47	2½
48	Plumbing, first part	46;48	3
49	Electrical work, first part	46;49	2
50	Joinery work, first section	46;50	1
51	Roof asphalt, third section	44, 45;51	½
52	Plumbing, second part	47, 48;52	3
53	Electrical work, second part	47, 48, 49;53	3
54	Erect partitions	33;54	2
55	Precast sills, first section	33;55	1
56	Exterior infill brickwork, second half	33;56	1
57	Precast sills, second section	55;57	1
58	In-situ sills, first section	55;58	1
59	Precast sills, third section	56, 57;59	1
60	In-situ sills, second section	58;60	1½
61	Windows, first section	58;61	1
62	In-situ sills, third section	59, 60;62	1
63	Windows, second section	61;63	1
64	Windows, third section	62, 63;64	1
65	Glazing, first section	61;65	1
66	Glazing, second section	62, 63, 65;66	1
67	Glazing, third section	64, 66;67	1
68	Plastering, first part	47, 48, 49, 50, 54;68	1
69	Plastering, second part	64, 68;69	½
70	Fix wall tiles	47, 48, 49, 50, 54;70	1

(continued)

Job number	Job	Job sequence	Job time (weeks)
71	Joinery work, second section	47, 48, 49, 50, 54 ;71	1
72	Asphalt flooring	69, 70 ;72	2
73	Install ceiling	71, 72 ;73	$1\frac{1}{2}$
74	Fix tile floors	71, 72 ;74	3
75	Joinery work, third section	71, 72 ;75	1
76	Decoration, first part	73, 74, 75 ;76	2
77	Test heating system	32 ;77	$\frac{1}{2}$
78	Decoration, second part and clean up	22, 32, 51, 52, 53, 67, 76 ;78	1

be started on that section while the piling of the second section is begun. The jobs continue through the project in series of steps, and this enables the project to be completed in the minimum time. The job chart drawn in Figure 11.8 shows the first feasible plan produced from the job sheet. It is obvious from this plan that the overall project duration of 30 weeks, indicated by the critical path of Jobs 1, 2, 4, 7, 10, 13, 16, 19, 23, 26, 29, 36, 38, 41, 46, 48, 68, 69, 72, 74, 76 and 78, can be maintained with reduced numbers of men employed on the site at certain times. This can be achieved by sequencing certain series, such as Jobs 40, 42, 43, 45, 44 and 51, or Jobs 54-67, while Job 22 and Jobs 32 and 77 could be left until later in the schedule. Job charts permit this sequencing to be done easily and show how movements of men between sites can be made most easily.

Project 19 — Building a reinforced concrete test laboratory

This project was required to be completed as quickly as possible from the date of award of the contract, and the completion date had to be stated in the contract, given its date of award. Figure 11.9 gives the job sheet for the project, in which the resource requirement for each job is quoted alongside the other relevant information. In view of the urgency of the job it was decided that all resources would be made available as and when they were required, with the restrictions that the contractors had only one excavator and had only four steelfixers on the strength. Obviously this is a case for job charts and a resource array.

The job chart for this project is shown in Figure 11.10.1. This is the first feasible plan for the project, and the critical path is seen to have a duration of 57 days, governed by Jobs 1, 3, 8, 15, 16, 17, 18, 19, 20, 21, 37, 38, 39 and 40. For the job times quoted, this is the minimum project duration, but it is necessary to determine whether or not this plan agrees with the resource restrictions. So a resource array is tabulated alongside the job chart, and in this array the total resource requirement for every day is given. We see that the plan is unacceptable on two counts. First, the jobs planned for Days 21, 22, 23 and 24 require the attention of five steelfixers, and secondly, the jobs planned

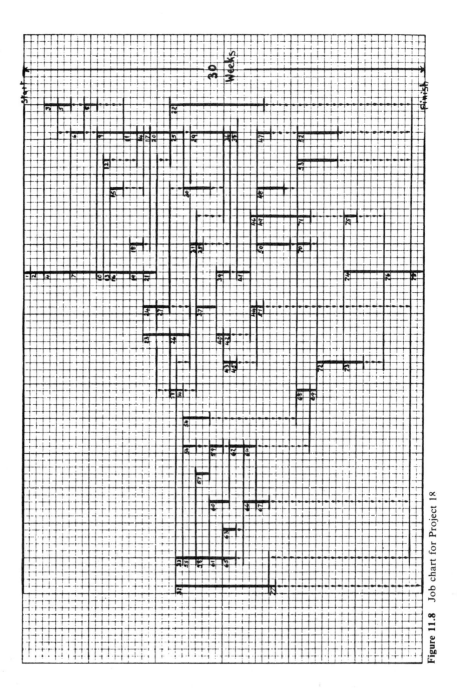

Figure 11.8   Job chart for Project 18

132

Figure 11.9 Job sheet for Project 19: building a reinforced concrete ??? laboratory

Job number	Job	Job sequence	Job time (days)	P	C	S	L	T	E
1	Accumulate equipment on site, build site shed	0;1	10	—	—				
2	Order and receive various piping	0;2	18	—	—				
3	Prepare site for laboratory	1;3	10						
4	Order and receive testing equipment	0;4	20						
5	Order and receive roof for laboratory	0;5	7						
6	Fabricate trench rebar	3;6	4	—	—	3	1	—	1
7	Fabricate trench forms	3;7	6	—	2	—	1	—	1
8	Make necessary site surveys	3;8	5						
9	Fabricate footing forms	3;9	7	—	2	—	1	—	1
10	Fabricate footing rebar	3;10	6	—	—	2	1	—	1
11	Order and receive hydraulic piping	0;11	20						
12	Make site excavation	8;12	3				1	—	1
13	Excavate footings	12;13	2	—	2	—	1	—	1
14	Place footing forms	9, 13;14	2	—	2	—	2	—	1
15	Excavate pipe trench	8;15	4	—	1	—	1	—	1
16	Form trench bottom	7, 15;16	3				2		
17	Place trench rebar	6, 16;17	2	—	—	2	1	—	—
18	Pour trench bottom	17;18	1				2	1	
19	Form trench walls	18;19	4	—	2	—	2	—	—
20	Place trench wall rebar	19;20	3			2	1		
21	Pour trench walls	20;21	1				2	1	

(continued overleaf)

Figure 11.9 (continued)

Job number	Job	Job sequence	Job time (days)	Resource requirement					
				P	C	S	L	T	E
22	Make up hydraulic piping as per work sheets	11;22	15	2	—	—	—	—	—
23	Install hydraulic test equipment	4;23	5	1	—	—	1	—	—
24	Check hydraulic test equipment	23;24	5	1	—	—	—	—	—
25	Test hydraulic piping	22, 24;25	5	2	—	2	—	—	—
26	Place footing rebar	10, 14;26	3	—	—	—	1	1	—
27	Pour footings	26;27	1	—	—	—	2	1	—
28	Erect wall forms	27;28	5	—	2	—	2	1	—
29	Place wall rebar	28;29	2	—	—	2	1	1	—
30	Set imbed metal	27;30	4	—	—	2	1	1	—
31	Pour walls	29, 30;31	1	—	—	—	2	1	—
32	Backfill footings	27;32	6	—	—	—	1	—	1
33	Install roof	5, 31;33	4	—	—	—	4	1	—
34	Clean, install and test hydraulic piping	25;34	10	2	—	—	2	1	—
35	Make up service piping as per work sheets	2;35	20	4	—	—	—	—	—
36	Test service piping	35;36	5	2	—	—	—	—	—
37	Clean and install service piping in trench	21, 36;37	4	2	—	—	2	1	—
38	Test installed service piping	37;38	3	2	—	—	—	—	—
39	Place trench covers	38;39	2	—	—	—	2	1	—
40	Check out and clean up	32, 33, 34, 39;40	5	—	—	—	2	—	—

P — Pipefitters   S — Steelfixers   T — Tower Crane
C — Carpenters   L — Labourers   E — Excavator

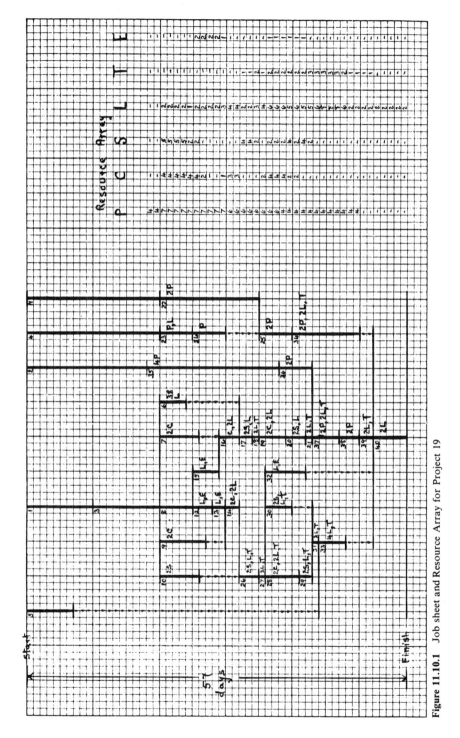

Figure 11.10.1 Job sheet and Resource Array for Project 19

135

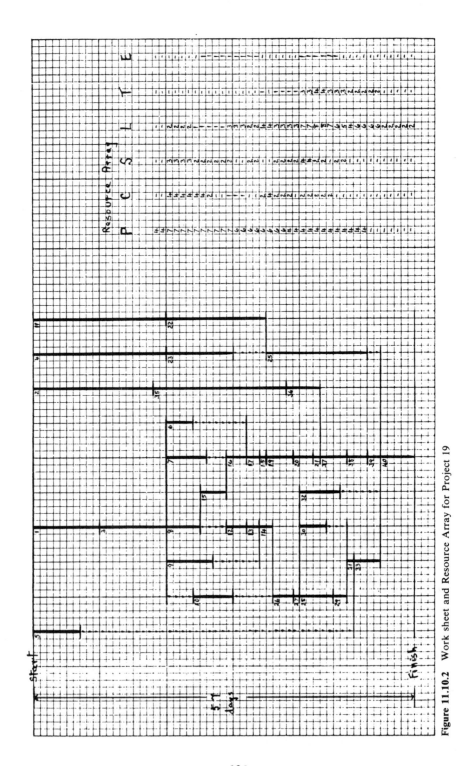

Figure 11.10.2 Work sheet and Resource Array for Project 19

for Days 26, 27, 28 and 29 require the use of two excavators. Fortunately for this particular project these two discrepancies can be resolved quite easily. The two jobs which require steelfixers on Days 21, 22, 23 and 24 are Jobs 6 and 10, and each has sufficient float to permit these two jobs to be sequenced. It does not matter which order of sequencing is used as far as this project is concerned, and the work chart in Figure 11.10.2 has assumed that the contractor would prefer to have three steelfixers on site for Days 21, 22, 23 and 24, after which one could be released for work elsewhere, and so Job 10 is scheduled to follow Job 6.

The requirement of two excavators for Days 26, 27, 28 and 29 is due to concurrent working of Jobs 12 and 13 with Job 15. This cannot be accepted, and since Job 15 lies on the critical path, consider moving Jobs 12 and 13 and see what happens. The sequence of jobs which starts with Job 12 ends with Job 33, after which there are 4 days' float before a time-line connects to the critical path. Thus, as Job 15 requires four days for completion, the Job 12-Job 33 sequence can be started 4 days later than shown in Figure 11.10.1, and this sequence of jobs can be completed in the time available without hindering any critical job. The final acceptable schedule is shown in Figure 11.10.2, and the resource array for this schedule is also shown. With transfer of the 4 day float from after Job 33 to before Job 12, Jobs 12, 13, 14, 26, 27, 28, 29, 31 and 33 have become critical.

The job chart permits resource scheduling to be done in such a way that the scheduler can see, at every move, exactly what is being done. The resource scheduling can be used, as in this project, to determine the daily requirements of the available resources, or to determine the project schedule for any given set of restrictions. If, for example, an additional restriction is imposed that only two carpenters can be made available at any one time, then from the work chart it is seen that Job 9 can be scheduled to follow Job 7; 1 day's interruption between Jobs 18 and 19 becomes necessary, and this, of course, extends the project duration by 1 day. Any other restriction, such as the availability of only one tower crane (which would mean sequencing Jobs 28, 29, 30, 31, 33, 34, 21, 37, 38, 39 and 40) or the availability of only five pipefitters would result in a considerable extension of the project duration.

Project 20 — Building an 11-storey block of flats

A short time ago Wates Limited contracted to build an eleven-storey block of flats in Bramley Hill, Croydon. The building was to be constructed from prefabricated sections, and it had been decided that all resources would be made available, as and when required, so that the building could be completed in the shortest possible time. When the site had been cleared and the foundations and base blocks laid, there were thousands of jobs relating to the structure which had to be completed, and these jobs were grouped into 49 compound jobs. Project 20 deals with these 49 compound jobs, and omits the preliminary work and the external works associated with the project; these were handled concurrently on another project.

The job sheet for the 49 compound jobs in Project 20 is given in Figure 11.11, and the job chart is shown in Figure 11.12. Most of the jobs in the job sheet are divisible into eleven parts, corresponding with the eleven storeys of the building, and the parts in each such job are identified by the subscripts '*a*' to '*k*'. These subscripts are not all written on the job chart, but the job lines are marked with dividers.

The job chart is drawn in the usual manner. Since many of the craftsmen follow each other up the building, floor by floor, there are many dependences

Figure 11.11 Job sheet for Project 20: building an 11-storey block of flats

Job number	Job		Job sequence	Job time (days)
1	R.C. frame typical	(11 parts)	0;1	22
2	Fix window frames	(11 parts)	1a;2	22
3	R.C. frame roof (first part)		1k;3	2
4	Brickwork to roof		3;4	2
5	R.C. frame roof (second part)		4;5	2
6	Fix window linings	(11 parts)	2a;6	22
7	Glazing externals	(11 parts)	2a;7	22
8	Fix door frames	(11 parts)	1b;8	22
9	Erect hoist (scaffold) (three parts)		1b;9a	2
			1f;9b	2
			1i;9c	2
10	Internal partitions	(11 parts)	9a;10a ⎫ 9b;10e ⎬ 9c;10i ⎭	22
11	Plumber, first fix	(11 parts)	1b;11	22
12	Basement steel partitions		1a;12	9
13	Chases for electricians	(11 parts)	10a;13	22
14	Electrical first fix	(11 parts)	10b;14	22
15	Gas first fix	(11 parts)	10b;15	22
16	Plasterer	(11 parts)	11a, 14a, 15a;16	22
17	Lifts	(11 parts)	10d;17	33
18	Floor screeds	(11 parts)	17d;18	22
19	Gas second fix	(11 parts)	16a, 18a;19	22
20	Carpenter, floors	(11 parts)	10a, 14a, 15a;20	22
21	Carpenter, second fix	(11 parts)	16a, 18a;21	22
22	Electrical, second fix	(11 parts)	21a;22	22
23	Plumber, second fix	(11 parts)	16a, 18a; 23	22
24	Wall tiling	(11 parts)	23a;24	22
25	Painter, internal	(11 parts)	24a;25	22
26	Floor tiles, flats	(11 parts)	25a;26	22
27	Carpenter, fittings, flats	(11 parts)	26a;27	22
28	Roof screed		5;28	3
29	Asphalter (to start 5 days after 28)		28;29	5
30	Asbestos tiles		29;30	3
31	Ventilation		29;31	6
32	Polystyrene to ceiling		29;32	6
33	Glazier, internal	(11 parts)	21f;33	11
34	Erect hanging cradles		29;34	1
35	Make good externally		34;35	5
36	Mastic pointing		35;36	3
37	Painting externally		36;37	6

(continued)

Job number	Job		Job sequence	Job time (days)
38	Lightning conductors (to start 3 days after 29)		29;38	2
39	Staircase balustrade	(11 parts)	1f;39	11
40	Plastering public areas	(11 parts)	29, 39;40	11
41	Grano landing, etc.	(11 parts)	40a;41	11
42	Floor screeds, public areas	(11 parts)	41a;42	11
43	Carpenter, second fix, public areas	(11 parts)	42f;43	6
44	Internal painting, public areas	(11 parts)	43a;44	8
45	Duradec, public areas	(11 parts)	44c;45	6
46	Floor tiles, public areas	(11 parts)	4 days of 45;46	3
47	Electrical fittings, public areas	(11 parts)	3 days of 45;47	4
48	Carpenter fittings, public areas	(11 parts)	4 days of 45;48	3
49	Dismantle cradles and clean up site		16, 17, 37;49	2

which could (and maybe which should) have been drawn in the chart as
ladders. For example, consider Jobs 23, 24, 25, 26 and 27, which must be
completed sequentially. Each part in any one of these jobs must be completed
before the next part in that job can be started; and each part of one job (such
as Jobs 23a) must be completed before the relevant part of the next job, (Job
24a) can be started. Instead of laddering all parts of these jobs, time lines have
been drawn linking the first and last dependences. In this job chart it was
convenient to draw job lines 40-48 moving to the left to reduce the chart
width. Instead of the customary float lines following job lines 30, 31, 32 and 38
to the end-line of the project, these job lines have been closed by gates. The
director arrow on the time line at the end of job line 29 indicates that both
Jobs 29 and 39 must be finished before Job 40 can be started, but that Jobs 30,
31, 32, 34 and 38 are dependent only on the completion of Job 29.

The job chart shows clearly the position in time of all the jobs, and shows
exactly how much float can be allowed any non-critical job or sequence of
jobs, so that such jobs can be manipulated in time to meet desired resource
allocation. The estimated duration of project 20 is 60 working days (ten 6-day
weeks) and the critical path is seen to consist of Jobs 1, 3, 4, 5, 28, the enforced
5-day wait, 29, 40, 41, 42, 43, 44, 45, 46, 47 and 48. But the sequence of Jobs
17, 18, 23, 24, 25, 26 and 27 has only 1 day float and in a project such as this
would be controlled as carefully as a critical sequence.

Project 21 — A paper mill project: changing the wire mesh

Operation of a paper mill is very like that of a chemical plant in that there is a
lot of maintenance work needed, and there is no installed spare equipment

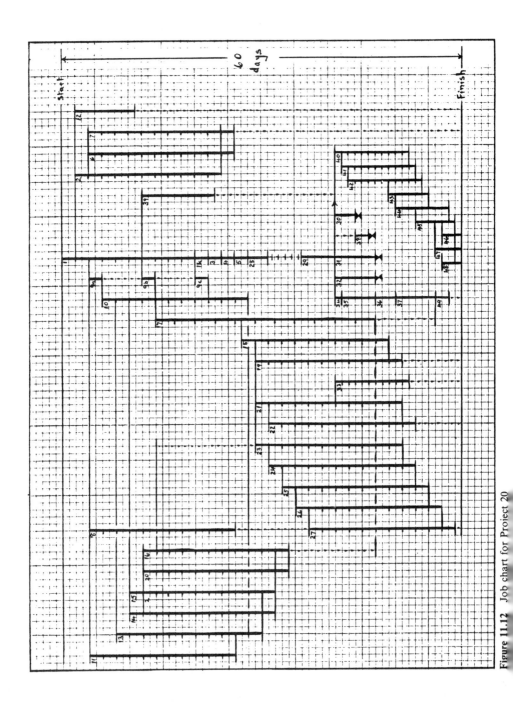

Figure 11.12   Job chart for Project 20

which permits the work to be done at convenient times. Many of the tasks recur regularly, and one of the most important of these is to change the wire mesh; this is a very complicated project which can be divided into a large number of job elements. In the following example many of the smaller jobs are not mentioned individually, but are included with their associated major jobs. The job sheet for this project is given in Figure 11.13 which includes only the job times for normal working. Since the work is performed as fast as possible and it is not feasible to bring any more workmen on to the site, there is no consideration of crash time to be made.

Figure 11.13 Job sheet for Project 21: changing the wire mesh

Job number	Job	Job sequence	Job time (min.)
1	Stop machine	0;1	5
2	Cut wire and remove in pieces	1;2	15
3	Remove suction boxes	1;3	15
4	Remove 16 table rolls	2;4	40
5	Remove shaker	1;5	35
6	Remove frame	3, 4;6	20
7	Remove spray pipes from breast roll	5;7	5
8	Remove strengthening bars and other spray pipes	6, 7;8	10
9	Remove doctor blades	8;9	5
10	Remove dandy rolls	8;10	10
11	Remove wash roll	8;11	10
12	Remove stretch roll	8;12	5
13	Lower breast roll	8;13	15
14	Remove dolly roll	8;14	5
15	Wash down couch roll. Remove connections and brackets	2;15	30
16	Bring new wire to couch roll	0;16	15
17	Slide new wire over couch roll	15, 16;17	25
18	Replace brackets and connections to couch roll	17;18	15
19	Unwind wire and remove supports	17;19	5
20	Extend wire over breast roll and insert table rolls	9-14, 18, 19;20	30
21	Replace breast roll and doctor blade	20;21	15
22	Replace stretch roll and doctor blade	20;22	10
23	Replace frame	21, 22;23	10
24	Replace dolly roll and shaker	23;24	10
25	Replace strengthening bars and spray pipes	23;25	10
26	Clean suction boxes	3;26	5
27	Replace suction boxes	23, 26;27	15
28	Replace table rolls	25;28	40
29	Replace dandy rolls	25;29	5
30	Run wire and adjust for tension	24, 27, 28, 29;30	10

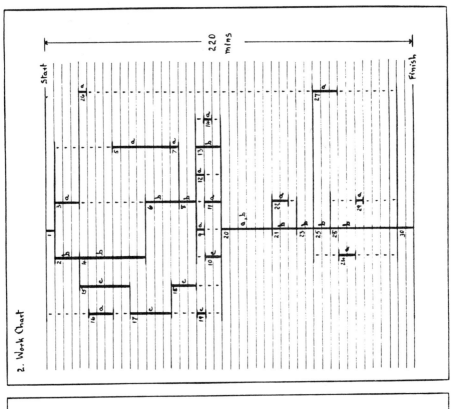

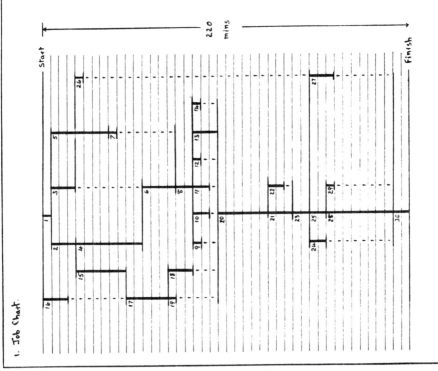

Figure 11.14 Job charts for Project 21

The job chart of Figure 11.14.1 is the first feasible plan for this project. It shows the critical path to consist of the twelve jobs 1, 2, 4, 6, 8, 13, 20, 21, 23, 25, 28 and 30, and also shows that every other job has some float, which may be of use in sequencing the jobs. The job chart shows that some jobs may be carried out concurrently, and in one instance it appears possible for all eight of Jobs 9, 10, 11, 12, 13, 14, 18 and 19 to be in progress at the same time.

However, one restraint must be enforced: only three teams of workmen are available, though one team can be split up to accommodate smaller jobs. It is necessary to sequence many of the jobs, therefore, to meet this restriction, and the work chart of Figure 11.14.2 shows how this can be done.

Job 16 can be started before Job 1 is complete, but it is not convenient to have to stop part-way through Job 16. Team 'a' attends to Jobs 3, 26, 16, 5 and 7; then splits to attend to Jobs 9 and 11 and Jobs 12 and 14; combines with team 'b' to attend to Job 20; and follows this by attending to Jobs 22, 27, 24 and 29. Of these jobs, Jobs 3, 26, 16, 5, 7, 9 and 11 are now critical; Job 22 retains its 5-minute float; and the sequence of Jobs 27, 24 and 29 has a float of 20 minutes. Team 'b' attends to all the jobs on the pre-determined critical path, while team 'c' attends to Jobs 15, 17, 18, 19 and 10, and all five of these jobs are now critical. The work chart shows that when the jobs are sequenced to meet manpower availability, only four, Jobs 22, 24, 27 and 29 remain non-critical.

Project 22 — Projects involving many departments: development of a new product

Each of the projects we have considered has been the concern of a single department or section of a company — usually that of the maintenance or of the construction department. But many occasions arise when a number of departments or sections of a company are required to undertake some activities in a project. As usual, the activities are inter-related, and each must be completed within certain time limits so that the project may achieve a satisfactory conclusion. Before committing itself to such a project, management must be informed of a number of factors, including the expected completion date and the expected participation of the various departments. Each department involved in the project wants to know when its particular activity can begin, when it must be completed, when it is expected to be completed (so that allocation of personnel may be made) and so on. Presentation of the data on a job chart enables everyone to see the project as a whole, and the degree of involvement of individual departments.

From time to time new products are developed in almost every industry, and during the period between the first suggestion that a product be developed and its eventual availability to customers, all departments of the company are likely to be involved in the development. The job sheet for one such project is given in Figure 11.15. Instead of the word 'chemical' in the descriptions of Jobs 2 and 7, similar job sheets might have 'metallurgical', 'physical', 'functional', or any other definitive term, and the job sheet is applicable to many industries. The job sheet shown is greatly simplified and omits mention of some departments, such as advertising, legal, and sales, which might be included in some projects. All jobs except Jobs 1, 15, 21 and 22

Figure 11.15   Job sheet for Project 22: development of a new product

Job number	Job	Job sequence	Job time (months)
1	Management decision to investigate possibilities	0;1	1
2	Initial chemical research	1;2	6
3	Initial market research	1;3	9
4	Investigate patent position	1;4	6
5	Investigate raw material supplies	1;5	6
6	Design pilot plant	2;6	4
7	Complete chemical process research	2;7	6
8	Construct pilot plant	3, 4, 6, 7;8	4
9	Manufacture in pilot plant	8;9	2
10	Package study	2;10	4
11	Storage tests (home, accelerated)	9, 10;11	4
12	Storage tests (overseas)	9, 10;12	12
13	Manufacture in pilot plant (continued)	9;13	2
14	Engineering evaluation for large scale manufacture	9;14	4
15	Management decision to proceed	11, 13, 14;15	1
16	Finalize raw materials contracts	5, 15;16	3
17	Full engineering design	15;17	8
18	Finalize package requirements	11, 12, 15;18	2
19	Erect manufacturing plant	17;19	12
20	Customer evaluation and field trials	13;20	12
21	Start manufacture	16, 18, 19;21	1 & on
22	Product available for dispatch	20, 21;22	1 & on

are compound jobs and could be subdivided into smaller elements if so desired; this subdivision would not be required for the plan to be presented to management.

The job chart for this project is shown in Figure 11.16 and this chart, with all other relevant data, is submitted to management for their consideration in Job 1. The job chart shows that the product is expected to be available for customers 45 months after the scheme is first submitted to management, though if required, smaller quantities can be made available from the pilot plant after 21 months. The critical path is seen to consist of Jobs 1, 2, 7, 8, 9, 11 and 14, 15, 17, 19 and one month of Job 21; all the other jobs have plenty of float and are not likely to become critical. The job chart shows clearly the time limits within which the activities of individual departments must be completed to avoid delaying completion of the project.

Project 23 — Projects involving many departments: implementation of a manpower planning system

An organisation which had a large head office and four manufacturing locations wished to introduce a system to act both as a manpower reporting

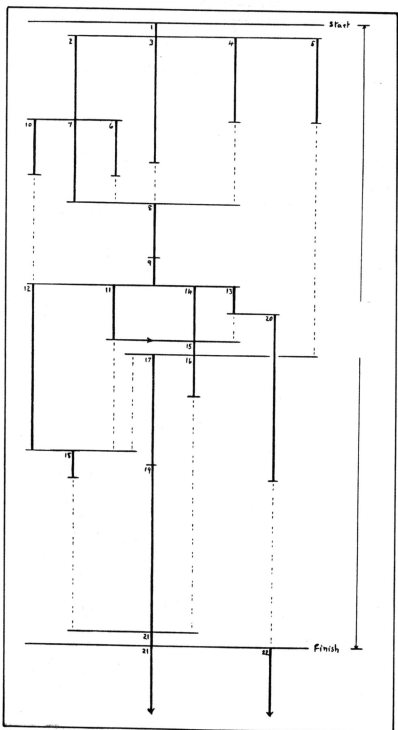

Figure 11.16 Job chart for Project 22

Job Number	Job	Job Sequence	Job Duration (weeks)
1	Discussions at Head Office	0: 1	2
2	Discussions with location representatives and feedback	1: 2	4
3	Data collection location 1	2: 3	2
4	Data collection location 2	2: 4	3
5	Data collection location 3	2: 5	3
6	Data collection location 4	2: 6	4
7	Data collection location 5	2: 7	8
8	Model formulation	1: 8	8
9	Model testing with given data	3,8: 9	4
10	Data checking	3,4,5,6,7: 10	2
11	Joint discussions, acceptance of system	9,10: 11	2
12	Punching data, computer run, inspection checks	11: 12	2
13	Discussion of results at Head Office	12: 13	2
14	Discussion of results at location 1	13: 14	2
15	Discussion of results at location 2	13: 15	2
16	Discussion of results at location 3	13: 16	2
17	Discussion of results at location 4	13: 17	2
18	Discussion of results at location 5	13: 18	2
19	Acceptance of plan by locations	14,15,16,17,18: 19	2
20	Final agreement at Head Office and implementation	19: 20	2

Figure 11.17 Job sheet for Project 23: implementation of a manpower planning system

system and a manpower planning system. The operational research department and the O&M department were involved in the project and compiled a job sheet which ran to several hundred jobs, of which the job sheet of Figure 11.17 is a greatly condensed version. The only constraint on the project was an instruction from the managing director that if there is to be a system it is to be working within six months.

The job sheet of Figure 11.17 shows that the longest time spent was on the operational research effort in providing and testing the model to be used (Jobs 8 and 9). The job chart of Figure 11.18.1 shows that as formulated the project cannot be completed in the desired 26 weeks, and it seems that a reduction in the duration of, at least, Job 7 is necessary. The durations of Jobs 3, 4, 5, 6 and 7 were in proportion to the number of employees at each location, assuming two persons on the data collection job at each location. Location 5 was asked to increase the effort and complete its data collection in five weeks, and the operational research department was asked to save one week between Jobs 8 and 9; actually the week was saved on Job 8. Figure 11.18.2 shows the work chart distributed to all concerned and the project was completed in 25 weeks.

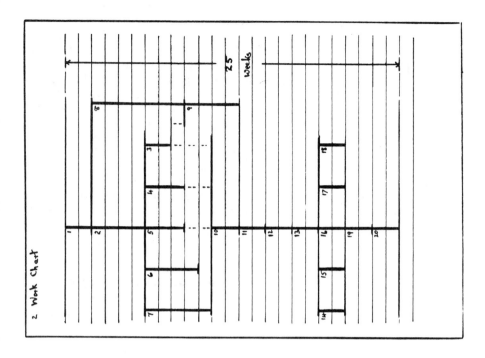

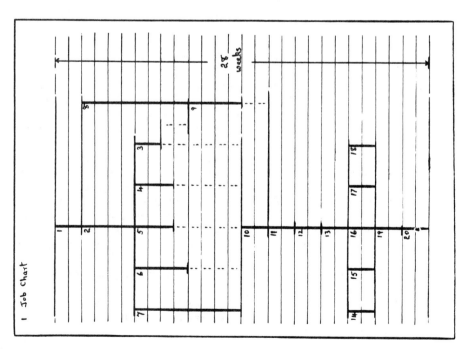

Figure 11.18 Job charts for Project 23

Project 24 — Planning: major project planning

In the chemical industry — and in most other industries too — a lot of work has to be done preparatory to running a new plant. The process has to be studied; equipment has to be studied and designed; materials and equipment have to be ordered, fabricated, and delivered; the site must be examined and approved; and everything has to be fitted together. Just as in plant maintenance, it is important that all activities are carried out at the appropriate times and in the correct sequence and that no delay is caused by unnecessary lack of momentum at any stage.

Figure 11.19 gives the master job sheet for the schedule of activities in the pre-production area of major installation projects in the chemical industry. Each job in this job sheet may be divided into a number (sometimes a very large number) of jobs which may be listed separately in the departmental job sheet. Usually any item of special significance is listed individually so that there is no chance of any slip over such item. The job sheet shows that with the available effort, 298 weeks work is involved in this major project planning, but the job chart of Figure 11.20 shows the schedule arranged so that the work is completed in one year. That is, as soon as management has approved the project, a maximum time of one year is available for all the project planning, ordering, fabrication, delivery, and installation. If the necessity arises additional effort is allocated at any stage to ensure that the one year is not exceeded. Figure 11.21 shows the modified chart, a planning work chart, issued to all personnel concerned in the project, though the charts issued are much larger in size and drawn in colours so that they are suitable for use as wall charts in controlling the project. The dates at the bottom of Figure 11.21 show the location in time of the first day of each month of the year, and it is normal practice to print finishing dates for each job on the chart.

Project 25 — Special Studies: stores procedure and inventory study

A large store held stocks of many items to provide an 'on-demand' supply service to several users. Existing procedures were ill-defined; high stocks of some items had been in store for a long time; demands on suppliers for immediate delivery of out-of-stock items were not unknown; and it was thought that a study based on sound inventory control principles would lead to big savings in capital, running costs, and smoother operation of the stores. The study was authorised and was undertaken by a small O&M team with assistance, as required, from an operational research scientist.

Figure 11.22 gives the job sheet for the project and shows 29 compound jobs, some of which consisted of a very large number of separate small jobs; the study team prepared a detailed list of planned activities which mentioned most of the jobs, though even here it was convenient to list jobs in groups. For example

Job 11 (Record by stock head for each item in the sample) was subdivided into

Average and maximum stock levels during the study period;
Longest lead-time in the last ten orders, or in the study period if less than ten items, (in weeks);

Job Number	Job	Job Sequence	Job Duration (weeks)
1	Prepare process control diagram and interpretation	0: 1	6
2	Prepare process flow and materials balance	0: 2	6
3	Prepare utilities and equipment information summary	0: 3	6
4	Study site and equipment arrangement	0: 4	9
5	Study model	0: 5	2
6	Prepare equipment layouts	4,5: 6	2
7	Prepare piping detail sheets	3: 7	7
8	Prepare engineering flow diagram	2: 8	7
9	Instrument design	1: 9	17
10	Electrical design	1,8:10	10
11	Prepare mechanical and piping drawings	6,7:11	8
12	Requisition and order major equipment	6,7:12	4
13	Mechanical design	6:13	6
14	Architectural design	6:14	10
15	Prepare foundation drawings	2 weeks of 14:15	6
16	Prepare steelwork drawings	2 weeks of 14:16	10
17	Ordering and delivery of foundation materials	15:17	7
18	Installation of structural and equipment foundations	17:18	6
19	Order, fabricate, and deliver structural steel	16:19	11
20	Requisition and order minor equipment	13:20	4
21	Fabricate and deliver minor equipment	20:21	18
22	Erect steelwork	18,19:22	5
23	Fabricate and deliver major equipment	12:23	22
24	Requisition and order piping	11:24	5
25	Deliver piping materials	24:25	9
26	Prepare electrical drawings	2 weeks of 10:26	10
27	Requisition and order electrical equipment	2 weeks of 26:27	4
28	Fabricate and deliver electrical equipment	27:28	18
29	Prepare instrument drawings	7 weeks of 9:29	16
30	Deliver electrical and instrument equipment	26,29:30	9
31	Requisition and order instruments	9 weeks of 9:31	11
32	Fabricate and deliver instruments	31:32	13
33	Prepare prefabricated piping	25:33	6
34	Set equipment	21,22,23:34	2
35	Install piping, electrics and instruments	28,30,32,33,34:35	6
36	Mechanical check-out and handover	35:36	5

Figure 11.19 Job sheet for Project 24: major project planning

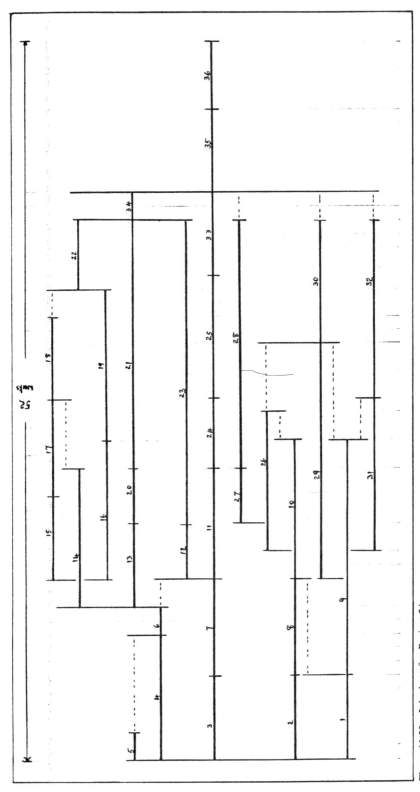

Figure 11.20 Job chart for Project 24

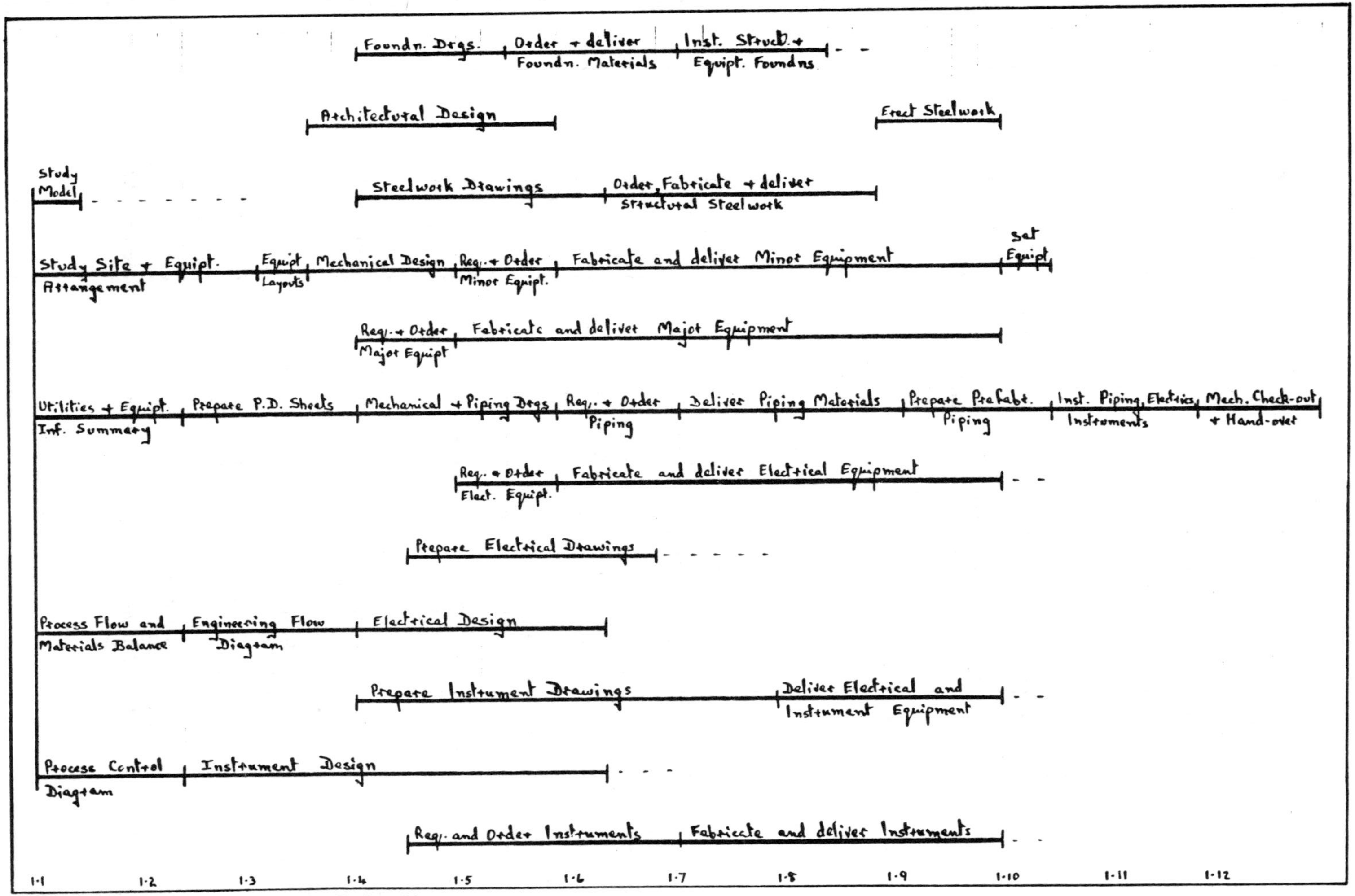

Figure 11.21 Planning work chart for Project 24

Job Number	Job	Job Sequence	Job Duration (half-weeks)
1	Decide reference period and obtain staff service details	0: 1	2
2	Summarise hours worked, etc., and cost at current rates	1: 2	1
3	Agree reference period costing with district accountant	2: 3	1
4	Obtain invoices for goods received during reference period	0: 4	1
5	Analyse invoices and record data	4: 5	6
6	Agree results with accountant and O.R.	5: 6	1
7	Record store layout showing area, volume, and types of storage for each stock head	0: 7	2
8	Draw plan of store	7: 8	1
9	Record current stock control, stores accounts, supplies and receipts, and issues procedures	3,7: 9	4
10	Operational research preliminary study	0:10	5
11	Determine number of separate items in each stockhead of sample	6,10:11	2
12	Record details by stockhead for each item in the sample	11:12	2
13	Calculate stock parameters	12:13	10
14	Calculate total value and total volume for these stocks	13:14	2
15	Agree calculations with accountant	14:15	1
16	Develop and agree new stock control system and prepare information flow sheet	9,15:16	5
17	Prepare procedure manual	16:17	9
18	Prepare new stores layout	8,17:18	4
19	List and cost new equipment needed; agree costs	18:19	2
20	Subdivide procedures into suitable elements	17:20	15
21	Estimate workload and required staff	20:21	4
22	Cost proposed staffing at current rates	21:22	2
23	Write appendices: "Present Situation"	19,22:23	2
24	Write appendices: "Discussion and Recommendations"	19,22:24	4
25	Write appendices: "Financial Implications"	19,22:25	2
26	Write details of the study	23,24,25:26	1
27	Assemble and check draft report	26:27	3
28	Agree draft report within department and with clients	27:28	4
29	Finalise and issue report	28:29	2

Figure 11.22 Job sheet for Project 25: stores procedure and inventory study

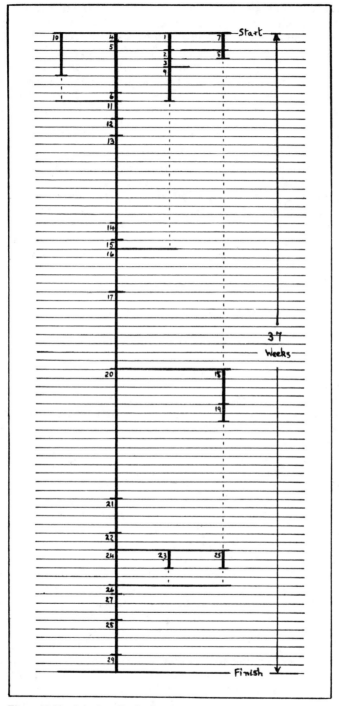

Figure 11.23   Job chart for Project 25

Average weekly demand; and Service level of safety factor.
Job 12 (Calculate for each item in the sample) was subdivided into:
   Average stock level; Maximum stock level; Re-order level; Re-order quantity; Value of average stock; Volume of average stock; Value of average stock in the study period. (Tabulate by stock heads.)
The job sequences shown in Figure 11.22 were derived by the study team, and other teams may show slight differences. Also, the job durations are those anticipated by the study team and are given in units of 'half-week'; if additional effort were made available, some of these job durations could be reduced.

Figure 11.23 shows the job chart for the project. Most of the jobs, 18 out of 29, lie on the critical path and the chart shows the overall duration of the project estimated at 37 weeks. The big items in the critical path are Job 13 (5 weeks), Job 17 ($4\frac{1}{2}$ weeks), and Job 20 ($7\frac{1}{2}$ weeks); and if additional effort were to be made available, these are the jobs on which this effort would be employed to show greatest reduction in the overall project duration. It was thought infeasible to arrange any of the jobs in the long chain Job 11 to Job 29 so that they could be carried out concurrently, and the project was run as shown on the job chart.

Project 26 — Special studies: planning system for operational research projects

Most large organisations have their own operational research departments while other organisations may open a contract for an operational research study to be made as and when required. When a study is to be made it is customary for the operational research team to prepare a schedule of the work thought to be involved in the study, so that the time involvement in various studies and the costs of these studies may be compared and effort directed in the line of maximum expected benefit. Many studies have similar format and work content, and Project 26 outlines a standard plan for these studies. In some respects Project 26 is very like project 24: each presents a standard work plan, but whereas Project 24 covers a large number of departments and disciplines, Project 26 is restricted to the personnel of one department.

Figure 11.24 gives the job sheet for Project 26 as prepared by the operational research scientists of one fairly large department, and some of the jobs listed may be worth comment. For example, the project is based on collection of four weeks' data (seven days data being processed in five days), but the data are checked for accuracy and conformity at the end of each week and are punched to provide computer input in two two-week batches. If a longer or shorter data collection period is necessary, then the durations of Jobs 18 to 27 must be amended to meet requirements. On completion of Job 10, or of Job 16, it may be necessary to reconstruct the model and re-test so that some repeat jobs, Jobs 6.1, 14.1, 15.1, 16.1 and 17.1 may need insertion in the plan. Any such circumstance will increase the overall project duration by the amount of repeat time spent on Job 6, and by the excess over the eleven days float spent on the other repeat jobs.

It will be noted that the job sheet does not include a job for 'implementation'. This is not a suitable place to present an argument whether implementa-

Job Number	Job	Job Sequence	Job Duration (days)
1	Receive and discuss request for assistance, and discuss and assess likely benefits	0: 1	4
2	Define the problem in clear and concise terms	1: 2	1
3	Draw a job sheet and job chart of the project	2: 3	1
4	Read relevant literature on the subject	2: 4	15
5	Interview all staff concerned in the project	2: 5	10
6	Build a model of the system	3,5,: 6	2
7	Determine data necessary to highlight problem areas	6: 7	2
8	Determine which data exist in a usable form	6: 8	1
9	Determine which data are to be collected, and in what form	7,8: 9	1
10	Prepare random sample of data	9:10	1
11	Determine the form of results required	9:11	2
12	Calculate sample size and estimate accuracy required	9:12	1
13	Design and print data collection forms	4,11,12:13	5
14	Prepare flowchart for computer program to analyse data	4,11,12:14	4
15	Write program	14:15	2
16	Punch program and test data	10,15:16	10
17	Computer run with test data. Analysis of output and assessment of model	16:17	5
18	Collect data (first week of four)	13:18	5
19	Examine first week's data for accuracy	18:19	2
20	Collect data (second week of four)	18:20	5
21	Examine second week's data for accuracy	20:21	2
22	Punch first two weeks' data	19,21:22	5
23	Collect data (third week of four)	20:23	5
24	Examine third week's data for accuracy	23:24	2
25	Collect data (fourth week of four)	23:25	5
26	Examine fourth week's data for accuracy	25:26	2
27	Punch second two weeks' data	24,26:27	5
28	Process data and examine output	17,22,27:28	10
29	Calculate accuracy of sample	28:29	1
30	Discuss results with staff involved. Assess benefits	29:30	4
31	Identify problem areas from computer output	28:31	5
32	Consider alternatives to system and test feasibility	30,31:32	10
33	Prepare draft report with recommendations	32:33	8
34	Hold discussions on draft report	33:34	7
35	Prepare and issue final report	34:35	6

Figure 11.24 Job sheet for Project 26: planning system for operational research projects

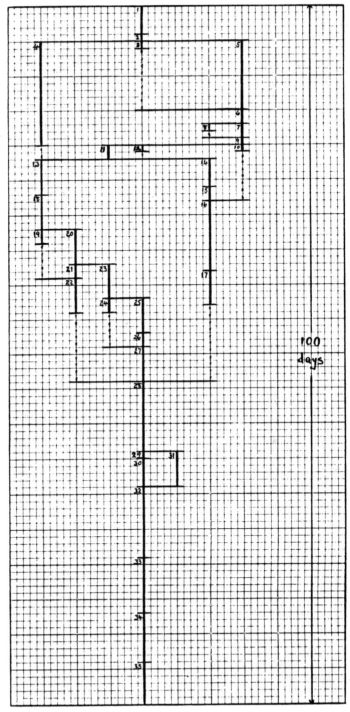

Figure 11.25 Job chart for Project 26

156

tion is or is not a function of the operational research scientist and a real part of the project, but it is usually the case that if the operational research scientist is to be involved in implementation, a new project assignment plan and job sheet etc, are prepared. Any further work done on a system, be it maintenance or modification, is considered as separate from the initial project, which relates solely to solution of a given problem and is concluded when recommendations are made for elimination or reduction of that problem.

Figure 11.25 shows the job chart for project 26, and it is noted that 22 of the 35 jobs are critical, while two other jobs, Jobs 8 and 12, are 1-day jobs with each having a 1-day float. The long chain of jobs, Job 25 to Job 35, occupies 58 days, and it has not been found possible to schedule the work to reduce this period. Note that splitting the data collection into four-weekly stints and two punching batches saved 11 days in the overall project duration.

Improbable uses of job charts

It has been pointed out to the author that the principle of job charts can be used to describe many situations and even to solve certain sorts of problems. This is so, but such uses of job charts, however numerous they may be, should not be considered unless some advantage is to be expected. Three examples are given in this section.

Project 27 — Inter-relationships of stages in a process

Any situation consisting of a number of separable components, which have some sequential relationships between them, can be described by a job chart. Figure 11.26 shows the inter-relationships of the processes involved in the manufacture of three chemicals. In this chart the lines numbered 1 to 22 represent stages in the manufacturing processes, and the length of each line indicates the total time cycle of each stage which is required to produce sufficient material for the next stage in the process. A stage-line may represent one batch, the total of several batches, or some length of run on a continuous unit. In the series shown the main product is obtained from stage 21 and cannot be expected until 60 hours after the last of the initial stages 1, 4 and 7 has started. Each of the other products, from stages 17 and 22, is available 5 hours earlier.

Project 28 — An allocation problem

Given that five tasks (a, b, c, d and e) can be performed by five people (A, B, C, D and E) in times shown by the array

	A	B	C	D	E
a	10	5	13	15	16
b	8	14	23	8	11
c	14	11	6	6	6
d	4	10	8	6	11
e	15	17	18	12	20

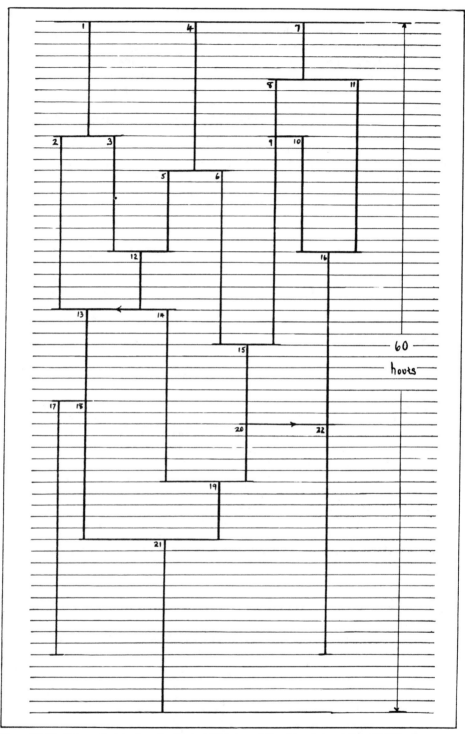

60
hours

Figure 11.26 Schematic job chart for Project 27: inter-relationship of stages in a process

158

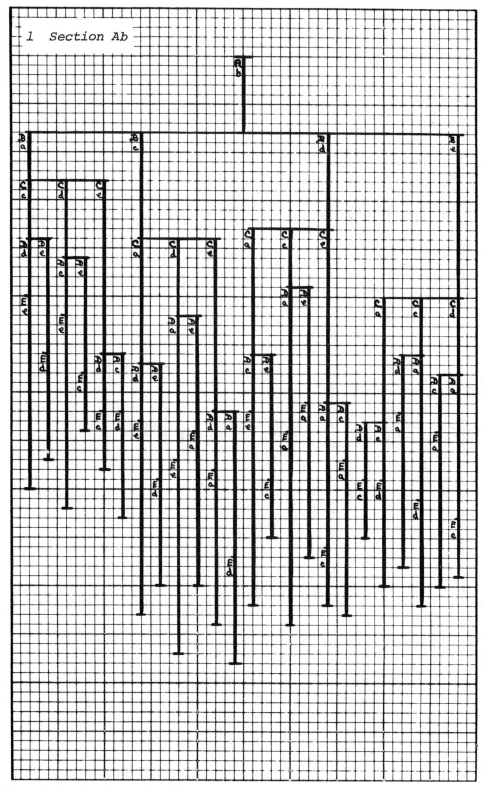

Figure 11.27 Allocation problem: Project 28   *(continued overleaf)*

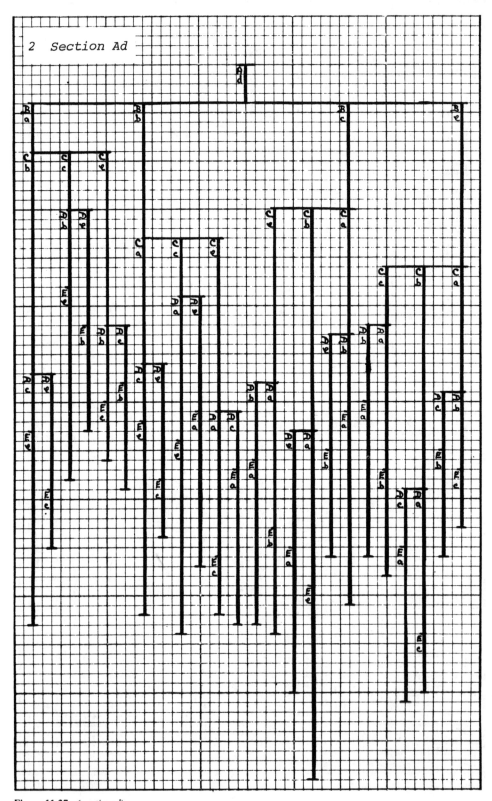

Figure 11.27 *(continued)*

it is required to determine that allocation of tasks among people to give the minimum total time. This can be found by drawing a series of charts, two of which are shown in Figure 11.27, Five such charts are needed to solve this problem, and these charts are nothing more than pictorial representations of all the possible allocations of tasks. The standard methods of solution os such problems are quite simple, and there can be no future for job charts in this field.

Project 29 — Shortest distance through a network

In Chapter 4 mention was made of finding the shortest distance through a network, and a network count procedure was described. Figure 11.28 shows the determination of the shortest route through the network of Figure 4.1 by the job chart technique. The procedure may be described as follows, noting that the arcs of the network may be traversed either way.

The line SS ruled across the page represents the starting point S, and from this line draw to scale the arcs $S,1 = 3; S,2 = 6$; and $S,3 = 5$, labelling the ends of these arcs 1.1, 2.1 and 3.1, respectively. From the point 1.1 draw the arc $1,4 = 5$, labelling the end of this arc 4.1. Note that there is no point in drawing the arc 4,2, as 4.1 is already further than 2.1 from SS, and the arc 4,2 cannot possibly lie on the shortest route through the network (though at present the arc 2,4 cannot be ruled out). From the point 4.1 draw the arcs $4,5 = 4$ and $4,7 = 8$, labelling the ends of these arcs 5.1 and 7.1 respectively.

From the point 2.1 draw the arcs $2,4 = 4$ and $2,5 = 6$, labelling the ends of these arcs 4.2 and 5.2 respectively; 4.2 is further than 4.1 from SS and so is discarded; 5.1 and 5.2 are the same distance from SS and are retained. There is no point in drawing the arc $2,3 = 3$, as 2.1 is already further than 3.1 from SS. From the point 3.1 we draw the arcs $3,2 = 3$; $3,5 = 4$; and $3,6 = 6$ labelling the ends of these arcs 2.2, 5.3 and 6.1, respectively; 2.2 is further than 2.1 from SS and is discarded; 5.3 is nearer than 5.1 and 5.2 to SS so that 5.1 and 5.2 are discarded. All three of the arcs leading from 2.1 have been rejected so point 2 cannot be on the shortest route through the network.

There is no point in drawing the arc $7,5 = 6$, as 7.1 is further than 5.3 from SS, so from the point 7.1 draw the arc $7,K = 3$, labelling the end of this arc as K.1. From the point 5.3 draw the arc $5,6 = 5$ to end in the point 6.2; 6.2 is further than 6.1 from SS and is discarded. From the point 6.1 draw the arc $6,K = 7$ to end in the point K.2. All possible traverses of all arcs have been considered; K.2 is nearer than K.1 to SS and is accepted as indicating the shortest route through the network, which Figure 11.28 shows to be S → 3 → 6 → K at a total distance of 16 units. This procedure is a graphical representation of the logic underlying the network count and may be useful for a preliminary demonstration but is thought not likely to be generally useful.

(Figure 11.28 overleaf)

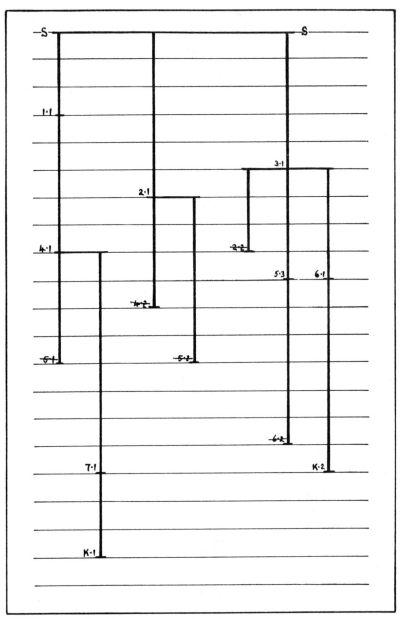

Figure 11.28 Project 29: shortest route through a network

Chapter 12

Management Objectives and Control

Various aspects of critical path analysis have been discussed and it has been shown how the available information may be processed to provide all the information necessary for efficient planning, scheduling and control of a project. The 'project' of critical path analysis should now be examined in a little more detail, using the six-point operational research plan to collate our information and knowledge.

1 Define the problem adequately

A procedure is required that will help in the planning, analysis and scheduling, and control of the project. The procedure must be applicable, immediately and without modification to meet different circumstances, to any and every project in which the total work that has to be done can be divided into distinct jobs, regardless of whether the project may be described as engineering, administrative, research, or anything else.

2 Identify the important factors

In order to deal with jobs then all the information must be available about each job that is necessary to place it in its correct location in the project. The technological sequence and the duration of each job are needed to start planning; any feasible alternative durations of each job are needed to start analysis; the resource requirement of each job and the cost of each job, under all feasible durations are needed to start scheduling. And for analysis and scheduling of projects which are to run concurrently, under the same control, a priority list is required. The important factors, therefore, are job sequence, job duration, job resource, job cost, and project priority.

3 Accumulate data

All the data relating to the sequence, duration, resource requirement and cost of every job in the project, and a priority list for multi-project work, must be available before planning, analysis and scheduling can be completed; and these data are presented most easily in a job sheet. The job sheet must contain all feasible alternatives and must be completely realistic with no built-in allowances for possible lapses. The greater part of the thinking through a project is done when preparing the job sheet, and this thinking is done by the planners or other members of the project team who have expert knowledge of the work to be done. It must be stressed that all the data relating to all the work to be done must be available; omission of jobs or part supply of data precludes planning.

4 Formulate a model

Two models have been considered that may be used for representation of a project: arrow networks and job charts. Also it has been seen that an arrow network must be subjected to arithmetic to determine the critical path (and this arithmetic may be performed on the job sheet), whereas the job chart shows the critical path immediately; a project and a three-project aggregate have been analysed and scheduled by means of job charts, and this could not be done from an arrow network. Additionally it has been seen that all the project information is with the project team. It seems reasonable that a model be provided which every member of the project team can understand and use. So the recommended model is a job chart, and the advantages of a job chart are reiterated as follows.

A job chart is extremely simple to construct. For someone who has never seen a job chart before, or who has never dealt in any way with critical path analysis, the training time required is of the order of a few minutes because a job chart can be drawn by anyone who can draw straight lines to scale; the only equipment required is a pencil, a ruler, and a sheet of ruled or squared paper. Once the job sheet is available, any member of the planning staff or project team can draw the job chart, and concurrent drawing of the job chart and thinking through the job sheet may be advantageous. There is no more difficulty in drawing a job chart for a big project with many jobs than there is in drawing a job chart for a small project with few jobs; one just works through the job sheet drawing the job lines as they are met. The job chart for the larger project is likely to require a larger sheet of paper, and probably the time required to construct a job chart increases proportionately with the number of jobs. A job chart carries only one series of numbers, that of the jobs listed in the job sheet, since every job line is numbered to correspond with the job it represents. Job durations are not marked on the job lines, and points of intersection of job lines do not exist. Time lines can be drawn across the chart, as required, to connect different parts of a project.

5 Assess the value of the model

A job chart is self-contained and self-explanatory; it presents a complete portrayal of a project and gives all the preliminary information about the project that is available at the time in an easily assimilable manner. Once a job chart has been drawn, the critical path or paths through the project are depicted clearly, and no arithmetic procedures or any other complications are involved. The job chart shows clearly the relationship in time of each job with every other job; it shows the duration of each job; and it shows the amount of float that may follow a non-critical job or sequence of jobs. The time scale used for the construction of the chart and details of job relationships quoted on the chart may vary with the intended use of the chart. Thus the planner, or scheduler, or any member of the project team can see his way through the project and can determine without difficulty any effects that may result from proposed alterations to the location in time of any critical or non-critical job. For very large projects job charts can be drawn for various sections of the project, and these can be related on a master job chart in which each job line is the critical path through the corresponding sub-project.

6 Implementation

A job chart can be used to simplify the planning, scheduling, and control stages of a project. In the planning stage the job chart is drawn as the first feasible plan for the project and can be used, firstly, to discuss critical jobs and any non-critical jobs which could become critical with small extensions to their durations, and any problems related to these jobs can be discussed and eliminated in advance or any special preparations can be determined; secondly, the possibilities attaching to the crashing of any job or sequence of jobs can be determined.

In the scheduling stage the job chart can be used to determine that pattern of job sequences which best fits the available resources. A resource array similar in construction to, and compiled with the aid of, the job chart can be constructed very easily and can be manipulated to satisfy any restrictions in resources that might be imposed on the project. The final schedule of jobs which describes the acceptable allocation of resources is called a work chart, and this may or may not be identical with the original job chart. This work chart is used by the project manager as a control chart.

In the control stage, which covers the actual work to be done in the project whether by a few men or by many men, the work chart can be used as a control guide so that the project manager ensures that all facilities are made available at the correct times. The main purpose of the work chart is to bring to attention those jobs whose durations control the overall project duration, so that effort and supervision can be concentrated on those jobs known in advance to be critical to the project. The result should be that all jobs are completed without unnecessary delays and the scheduled project completion date achieved. If, through any uncontrollable cause, extension of a critical job duration occurs, or the duration of a non-critical job is extended so that it becomes a critical job and perhaps produces a new critical path through the project, the work chart can be altered very easily. The common practice is to cut across the chart and insert a piece of paper of the right length to cover the extension.

Also, the work chart can be used as a progress chart to record the amount of work completed so that at the end of any day (or week, or month) the project manager knows exactly the status of the project. This can be done in several ways, as for example: the job lines can be cross-hatched to show how much work has been completed on each job, or a coloured line can be drawn alongside the job line. In this case the work chart may need to be drawn with a different scale unit from the job chart used for planning and scheduling the project; whereas the planner might be content with showing a job duration as two weeks, the project manager might want that particular job represented by a job line divisible into ten-day units suitable for a five-day working week.

When the direct costs of completing each job in a project are known and the possibility of crash working is to be considered, a job chart enables a quick determination to be made of which jobs in a project should be crashed and what benefit is likely to accrue from the crashing. This permits the maximum reduction in the project duration to be made for the minimum additional expenditure of money, if it can be shown that it is profitable to make the saving in time. Where the pursuance of any project involves a loss of production which may be evaluated as a loss of profit per day (or per whatever time unit is used) the benefit to be achieved by crashing the project, or by crashing any critical job or series of jobs, can be obtained by a straightforward comparison of the premium cost of the crashing with the saving in production value over the crash time period; use of a job chart greatly simplifies this comparison. If indirect costs, or overheads, are to be included in the project costing, a minimum total project cost can be determined; with a job chart it is simple to determine at any time how much of the budgeted expenditure has been spent and how progress of the project compares with budget.

Where a number of projects are being run at the same time job charts enable the project manager to allocate the available resources to satisfy the requirements of every job in the various projects. This is done easily by compounding the resource arrays of the individual projects and then manipulating the master array so that the schedules of jobs give minimum project durations compatible with the resource restrictions. Once this is done it is possible to compile a detailed work schedule, with enforced intervals and floats, for each item in the resource list; this procedure is especially useful for specialist workmen or for specialised items of equipment which move from one project to another, taking part maybe in only one or two jobs in each project.

Job charts do not permit the inclusion of any uncertainty factors. They are constructed on the basis of one time estimate for each job, and this estimate is the best estimate that is available. Uncertainty factors are a purely fictional contrivance which cannot be applied to job charts and which are not required because job charts are used to remove as many uncertainties as possible from the project, to give the project manager a feasible, reasonable, and fixed schedule of work which management expects to be completed by the agreed finishing date.

Job charts have been used successfully for several years, with projects large and small, where the requirements are that one is to be able to see one's way through the project and to be able to manipulate the job elements and the resource availabilities. Everyone can use them; they work and can be seen to work.

Chapter 13

Example Projects for Practice

This chapter contains twelve example projects which may be used by the reader to try the procedures described in earlier chapters of this book and assure himself that they really work. Some of the planning charts and sheets for these example projects are included, and some arrow networks have been drawn, but the purpose is for the reader to work through these projects for himself and acquire all the expertise he wants. The job charts are quoted as prepared originally by the works engineers and planners.

Examples X.1-X.4 are small projects which may be used as examples to determine any of a variety of planning requirements such as cost and layout of the minimum time project, time and layout of the project for some agreed cost, or comparison of cheapness ratings against a given project value. Examples X.5-X.8 may be used as the previous examples and may be used for scheduling manpower. These increase in complexity from the single project of Examples X.5 and X.6 to the double project of example X.7 and the triple project of Example X.8. The reader may prepare full planning documentation under various expenditure-profit relationships. Examples X.9-X.12 refer to somewhat larger single projects. Only the job sheets are given, and the reader is invited to prepare full planning documentation. Manpower requirements for the jobs in these projects may be written in by comparison with the jobs in Examples X.5-X.8, and scheduling may be done under various availability assumptions.

Example X.5 — Project for scheduling manpower

Figure 13.25 gives the job sheet for Example X.5, which is to be scheduled under the following restraints:
1 The maximum numbers of men that can be assigned to the project on any day are

2 Carpenters	3 Plumbers
3 Electricians	1 Rigger
10 Fitters	1 Welder

2   To avoid undue interference with other projects craftsmen are required to be engaged continuously on this project. That is, there should be no undue variation in day-to-day craft demands.

3   Labourers and other workers are available as required and so need not be scheduled.

Figure 13.26 shows the job chart and the resource array to match this job chart, and Figure 13.27 shows a first attempt at a work chart with its associated resource array. Figure 13.28 shows the arrow network for the project.

Example X.6 — Project for scheduling manpower

Figure 13.29 gives the job sheet for Example X.6 which is to be scheduled under the following restraints:

1   The maximum number of men that can be assigned to the project on any day are:

      2 Plumbers   2 Riggers   1 Welder

2   Other craftsmen can be supplied as required, but the site cannot accommodate more than 15 men at a time.

Figure 13.30 shows the job chart and the resource array to match this job chart, and Figure 13.31 shows a first attempt at a work chart with its associated resource array. Figure 13.32 shows the arrow network for the project.

Example X.7 — Joint operation of two projects

Figures 13.33 and 13.34 show the job sheets prepared for two projects which are to be completed by one engineering team under the control of the divisional engineer. After discussion of the projects with the works management, it was decided that each project was to be completed within the time limit shown on the work sheet, if possible, as there was very little extra cash available for crash working; completion of the projects was important but at standard costs.

Resource restrictions referred only to manpower and the team was restricted to

      4 Bricklayers   4 Plumbers
      8 Carpenters   3 Riggers
      2 Electricians   2 Welders
      30 Fitters

and whereas work was available for any of these craftsmen not engaged on the projects, it was thought that the run of work and utilisation of craftsmen should be as smooth as possible.

Figure 13.35 shows the combined job chart and manpower array for the two projects, and it is necessary to manipulate the job lines to meet the restraints listed above. Figure 13.36 shows a first attempt at preparation of a work chart, which shows one job crashed at an additional expenditure of £150. Figures 13.37 and 13.38 show the project cost-duration tabulations, and Figures 13.39 and 13.40 show the stages of crashing the projects. Figures 13.41 and 13.42 shows arrow networks for the two projects.

Example X.8 — Installation of manufacturing plant

For convenience of control, the installation of new manufacturing equipment, all of which was to be placed in and around one building, was divided into three projects. Figures 13.43, 13.44 and 13.45 show the job sheets prepared for the three projects, and it is noted that there are several dependences between projects. The three projects were to be completed by one engineering team under the control of the divisional engineer; though because of the importance of the overall project, arrangements could be made for additional craftsmen to be made available at high premium costs.

The original job charts and manpower array are shown in Figure 13.46, from which it is required to prepare the details required at the planning stage:

1 Examine the feasibility of completing the installation in 42 days with the availability of craftsmen as

	Engineering team	Additional availability
Carpenters	4	3
Electricians	3	3
Fitters	20	8
Plumbers	4	1
Riggers	2	2
Welders	3	1

2 Determine the schedule and cost of completing the installation in minimum time with the above availability of craftsmen and the value of the products to be made estimated at £5,000 per day.

3 Determine the overall duration for the complete installation, and compare costs with loss of profit, if manpower availability is restricted to members of the engineering team.

Job Number	Description	Job Sequence	Standard		Crash	
			Duration (days)	Cost (£)	Duration (days)	Cost (£)
1	Lead time for management approval	0:1	3	–	1	–
2	Line availability	0:2	4	–	4	–
3	Measure and sketch	1:3	2	60	1	80
4	Develop materials list	3:4	1	20	1	20
5	Procure pipe and flanges	4:5	5	170	3	220
6	Procure valves	4:6	9	60	5	120
7	Prefabricate sections in workshop	5:7	5	240	2	400
8	De-activate line	2,4:8	1	20	1	20
9	Erect scaffold	4:9	2	60	1	100
10	Remove old pipe and remove to scrapyard	8,9:10	4	80	2	200
11	Place new pipe in position	7,10:11	6	300	3	600
12	Weld pipes	11:12	2	20	1	60
13	Place valves in position	6,8,9:13	2	20	1	50
14	Fit up valves and pipes	12,13:14	2	20	1	50
15	Pressure test on system	16:15	2	10	1	20
16	Insulate	12,13:16	4	80	2	140
17	Remove scaffold	14,16:17	2	20	1	50
18	Clean up	15,17:18	2	20	1	50
Project		?	?	£1200	?	?

Figure 13.1 Job Sheet for Example X.1

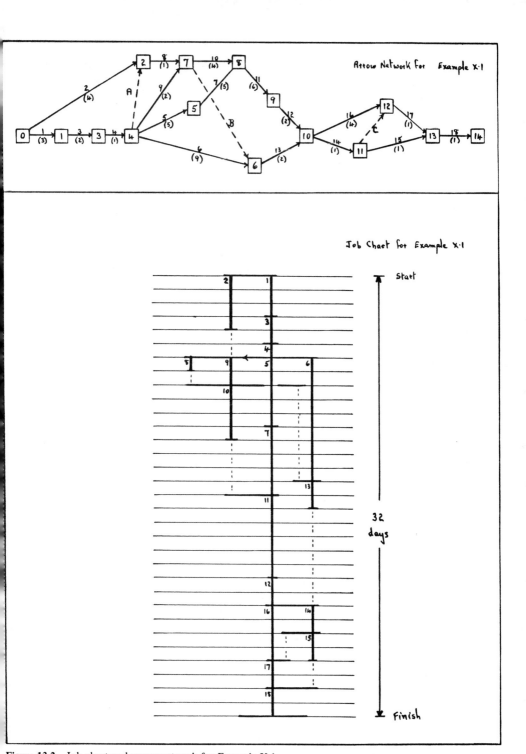

Figure 13.2 Job chart and arrow network for Example X.1

171

Crash Job	Days Saved	Premium Expenditure	Cheapness Rating
1	2	£ -	-
5	2	50	25
18	1	30	30
16	2	60	30
12	1	40	40
15 and 17	1	40	40
7 and 9	3	200	66·6
6 and 11	3	360	120
Project	15	780	52

Figure 13.3 Evaluation of job crashing of Example X.1

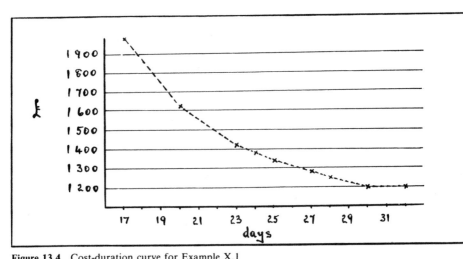

Figure 13.4 Cost-duration curve for Example X.1

Day	Concurrent Jobs	Rating
1	2	1.10
2	2	1.10
3	2	1.10
4	2	1.10
5	1	0.55
6	1	0.55
7	4	2.21
8	3	1.66

Day	Concurrent Jobs	Rating
9	3	1.66
10	3	1.66
11	3	1.66
12	3	1.66
13	2	1.10
14	2	1.10
15	2	1.10
16	2	1.10

Day	Concurrent Jobs	Rating
17	2	1.10
18	1	0.55
19	1	0.55
20	1	0.55
21	1	0.55
22	1	0.55
23	1	0.55
24	1	0.55

Day	Concurrent Jobs	Rating
25	2	1.10
26	2	1.10
27	2	1.10
28	2	1.10
29	1	0.55
30	1	0.55
31	1	0.55
32	1	0.55

Figure 13.5   Congestion ratings for the job chart of Figure 13.2

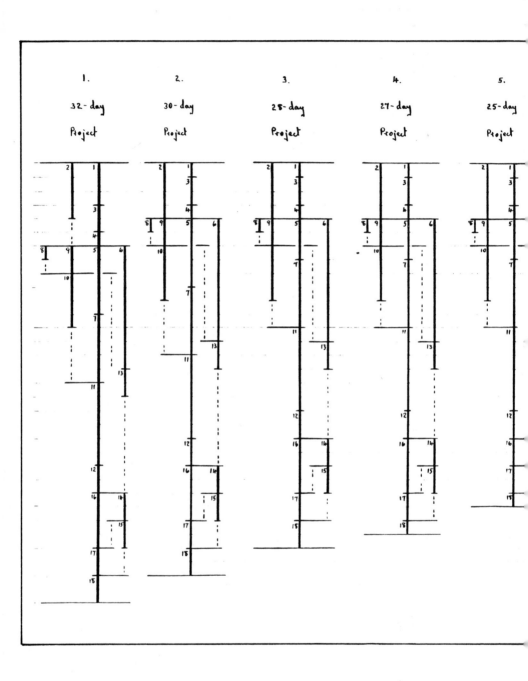

1. 32-day Project

2. 30-day Project

3. 25-day Project

4. 27-day Project

5. 25-day Project

Figure 13.6 Project cras

174

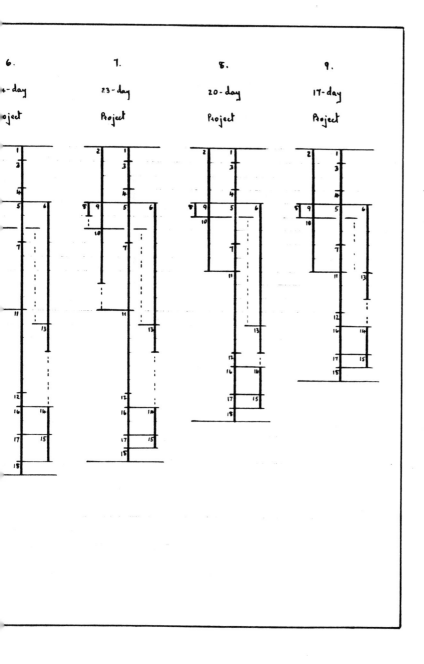

6.
...-day
...oject

7.
23-day
Project

8.
20-day
Project

9.
17-day
Project

...required by Figure 13.3

Job Number	Job Sequence	Standard Duration	Cost	Crash Duration	Cost
1	0: 1	6 days	£ 700	3 days	£ 1 600
2	0: 2	3	400	3	400
3	2: 3	4	800	3	1 100
4	2: 4	5	1 100	4	1 400
5	0: 5	5	900	3	1 500
6	5: 6	9	1 350	6	2 100
7	3: 7	3	470	2	700
8	1, 7: 8	4	750	2	1 650
9	3: 9	4	650	2	1 500
10	4: 10	5	800	3	1 500
11	4: 11	3	450	3	450
12	9, 10: 12	3	500	2	900
13	9, 10: 13	3	520	2	900
14	8, 12: 14	7	1 000	5	1 400
15	11: 15	7	980	4	1 700
16	13, 15: 16	3	500	2	900
17	11: 17	5	850	4	1 100
18	6: 18	5	900	3	1 750
19	6: 19	6	830	3	1 900
20	14, 16, 17, 18, 19: 20	3	420	2	750
21	14, 16, 17, 18, 19: 21	5	630	3	1 000
22	20, 21: 22	3	400	2	650
Project		?	15 900	?	?

Figure 13.7   Job sheet for Example X.2

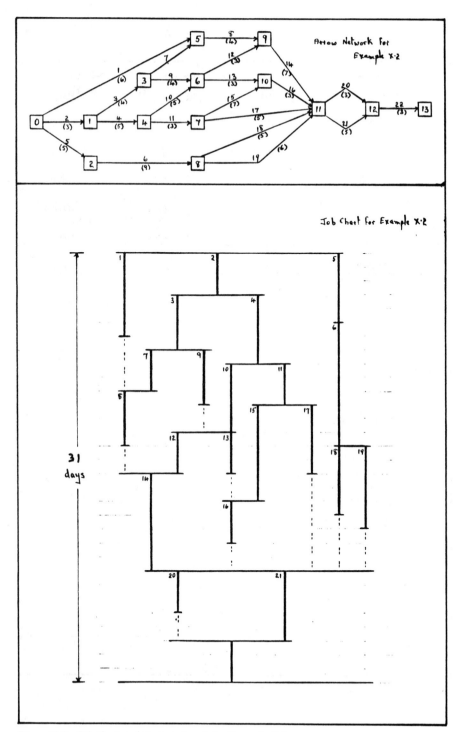

Figure 13.8 Job chart and arrow network for Example X.2

177

Crash Jobs	Days Saved	Premium Expenditure	Cheapness Rating
21	2	£ 370	185
14	2	400	200
22	1	250	250
4	1	300	300
6, 8, 9, 10, 12 and 15	3	4 320	1 440
Project	9	5 640	626·6

Figure 13.9 Evaluation of job crashing of Example X.2

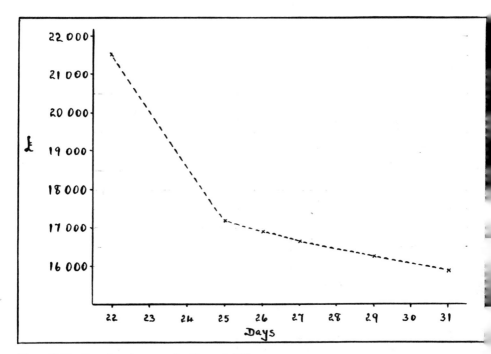

Figure 13.10 Cost-duration curve for Example X.2

Day	Concurrent Jobs	Rating
1	3	0.77
2	3	0.77
3	3	0.77
4	4	1.02
5	4	1.02
6	4	1.02
7	3	0.77

Day	Concurrent Jobs	Rating
8	5	1.28
9	5	1.28
10	4	1.02
11	6	1.53
12	7	1.80
13	6	1.53
14	5	1.28

Day	Concurrent Jobs	Rating
15	5	1.28
16	4	1.02
17	7	1.80
18	2	0.51
19	2	0.51
20	2	0.51
21	1	0.26
22	1	0.26

Figure 13.11   Congestion ratings for job chart 6 of Figure 13.12

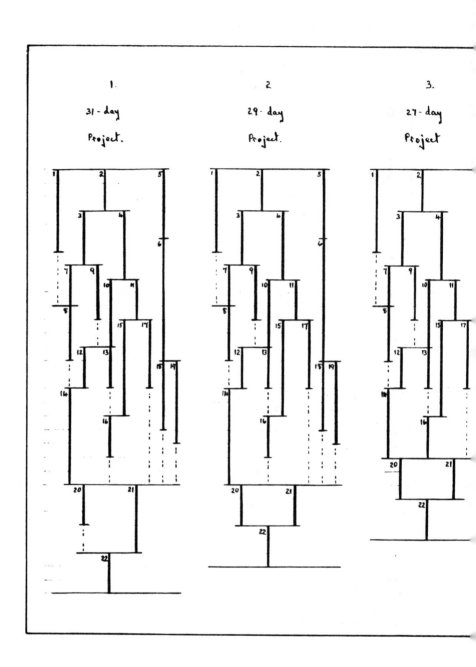

Figure 13.12 Project cra

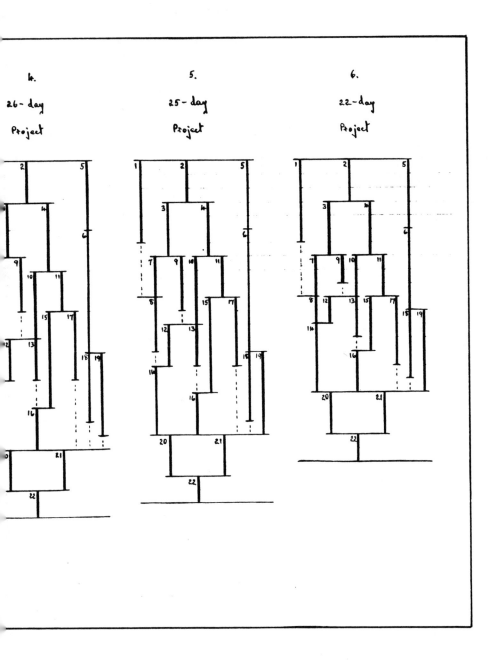

4.

26 - day

Project

5.

25 - day

Project

6.

22 - day

Project

...equired by Figure 13.9

Job	Job Sequence	Standard Duration	Standard Cost	Crash Duration	Crash Cost
1	0 : 1	4 days	£ 580	2 days	£ 1 300
2	1 : 2	3	420	2	750
3	1 : 3	5	900	3	1 400
4	1 : 4	4	650	2	1 400
5	2 : 5	4	600	3	800
6	0 : 6	3	480	2	800
7	0 : 7	2	400	2	400
8	6 : 8	3	500	2	850
9	8 : 9	3	410	2	680
10	7, 6 : 10	4	860	3	1 000
11	7, 6 : 11	5	1 000	3	1 700
12	9, 10, 11 : 12	3	480	2	850
13	9, 10, 11 : 13	4	720	3	1 100
14	12 : 14	4	670	2	1 500
15	3, 4, 8 : 15	5	990	3	1 800
16	5 : 16	3	430	3	430
17	5 : 17	5	1 100	3	1 650
18	12 : 18	2	370	2	370
19	13, 14, 15, 16, 18 : 19	4	620	2	1 400
20	13, 14, 15, 16, 18 : 20	6	1 500	3	3 550
21	13, 14, 15, 16, 18 : 21	4	840	2	1 900
22	21 : 22	3	510	3	510
23	22 : 23	3	500	2	840
24	17, 19 : 24	4	660	2	1 500
25	20, 23, 24 : 25	3	450	2	720
26	20, 23, 24 : 26	6	900	3	2 750
27	25, 26 : 27	4	750	2	1 850
Project		?	18 280	?	?

Figure 13.13   Job sheet for Example X.3

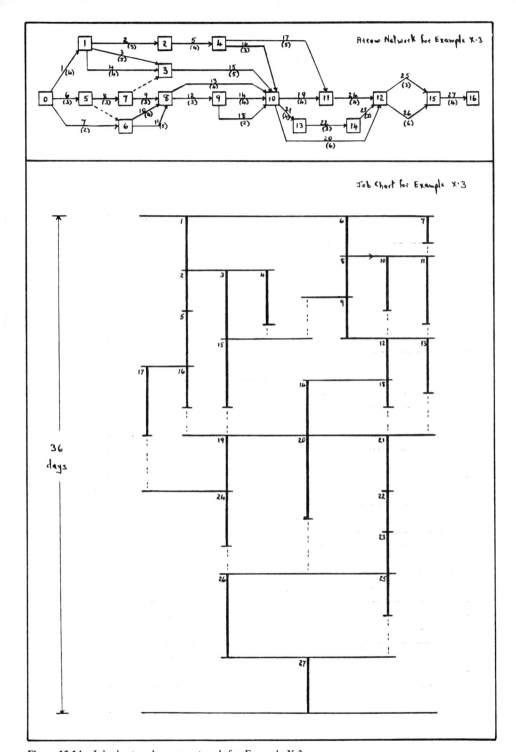

Figure 13.14 Job chart and arrow network for Example X.3

183

Crash Jobs	Days Saved	Premium Expenditure	Cheapness Rating
9	1	270	270
6	1	320	320
23	1	340	340
27	2	1 100	550
26	3	1 850	616·6
1 and 14	2	1 550	775
19 and 21	2	1 840	920
2, 5, 8, 11, 12 and 15	2	2 760	1 380
Project	14	10 030	716·4

Figure 13.15 Evaluation of job crashing of Example X.3

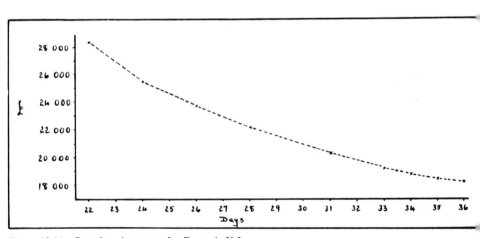

Figure 13.16 Cost-duration curve for Example X.3

Day	Concurrent Jobs	Rating
1	3	0·84
2	3	0·84
3	6	1·67
4	6	1·67
5	6	1·67
6	5	1·40
7	4	1·12

Day	Concurrent Jobs	Rating
8	5	1·40
9	6	1·67
10	6	1·67
11	4	1·12
12	4	1·12
13	3	0·84
14	3	0·84

Day	Concurrent Jobs	Rating
15	3	0·84
16	3	0·84
17	1	0·28
18	2	0·56
19	2	0·56
20	2	0·56
21	1	0·28
22	1	0·28

Figure 13.17 Congestion ratings for job chart 4 of Figure 13.18

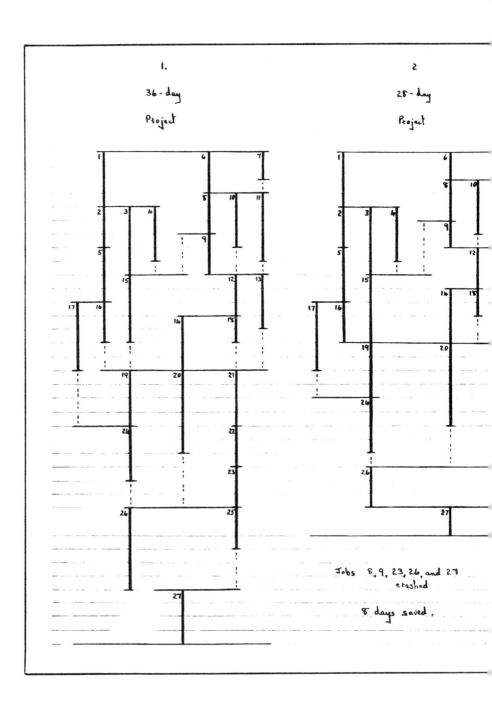

1.

36-day

Project

2

25-day

Project

Jobs 8, 9, 23, 26, and 27
crashed

8 days saved.

Figure 13.18 Project cr

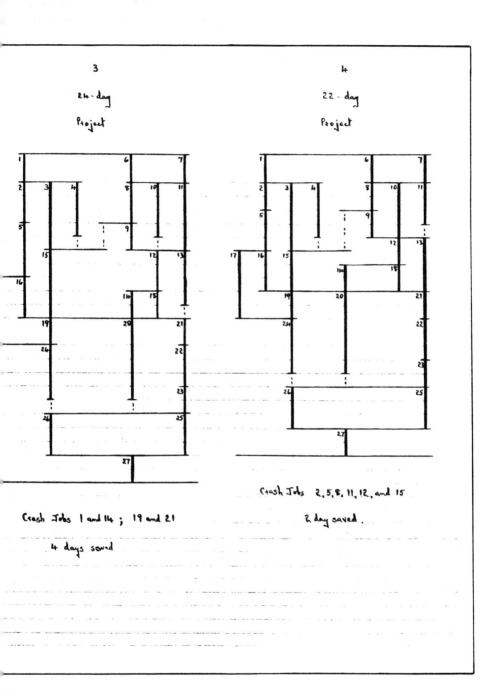

3

24-day

Project

4

22-day

Project

Crash Jobs 2,5,8,11,12, and 15

2 day saved.

Crash Jobs 1 and 14 ; 19 and 21

4 days saved

s required by Figure 13.15

Job Number	Job Sequence	Standard		Crash	
		Duration	Cost	Duration	Cost
1	0 : 1	4 days	£ 800	2 days	£ 2 000
2	0 : 2	4	1 200	4	1 200
3	0 : 3	4	960	3	1 350
4	1 : 4	3	1 300	2	2 050
5	1 : 5	4	820	2	1 800
6	1 : 6	4	1 000	3	1 350
7	2 : 7	5	1 050	3	2 000
8	2 : 8	6	1 500	4	2 500
9	3 : 9	3	900	2	1 350
10	3 : 10	4	1 100	2	2 450
11	10 : 11	3	750	2	1 200
12	4 : 12	5	1 200	3	1 950
13	5, 6, 8 : 13	3	810	3	810
14	5, 6, 8 : 14	5	1 400	4	1 750
15	7, 9 : 15	3	900	2	1 400
16	13, 15 : 16	2	600	2	600
17	4, 11 : 17	6	3 600	4	6 300
18	4, 11 : 18	4	1 820	3	2 500
19	12, 14 : 19	4	1 600	2	4 000
20	12, 14 : 20	5	1 350	4	1 900
21	18 : 21	4	960	4	960
22	16, 17 : 22	3	840	3	840
23	19, 20 : 23	3	750	2	1 050
24	21, 22 : 24	4	1 000	2	2 500
25	21, 22 : 25	6	1 200	3	3 000
26	19 : 26	5	950	3	1 500
27	26 : 27	4	800	2	2 000
28	26 : 28	4	880	3	1 200
29	23, 24 : 29	3	720	2	1 000
30	28, 29 : 30	3	600	2	800
31	25, 29 : 31	4	600	2	1 200
32	27, 30, 31 : 32	3	600	2	800
	Project	?	£ 34 560	?	?

Figure 13.19   Job sheet for Example X.4

188

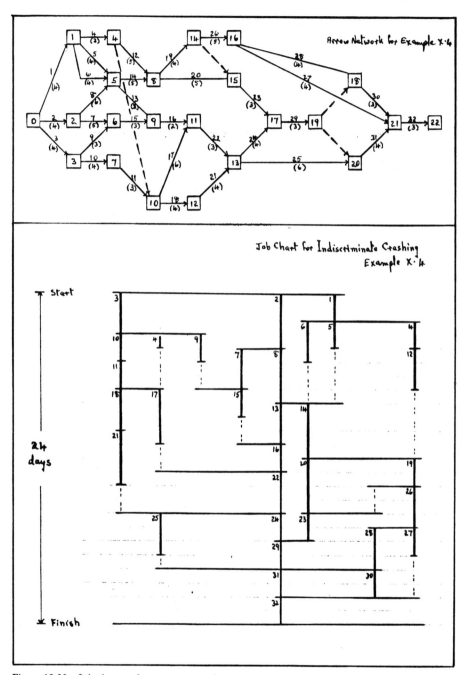

Figure 13.20 Job chart and arrow network for Example X.4

Crash Jobs	Days Saved	Premium Expenditure	Cheapness Rating
32	1	£ 200	200
28, 30, 31	2	1 120	560
3, 8, 11	2	1 830	915
7, 10, 23, 26	2	3 150	1 575
14, 19, 20, 24, 25, 29	3	6 880	2 293
Project	10	13 180	1 318

Figure 13.21 Evaluation of job crashing of Example X.4

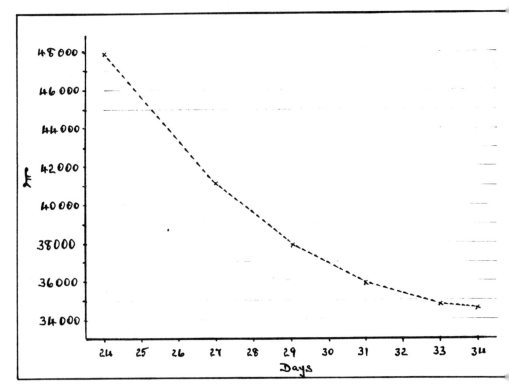

Figure 13.22 Cost-duration curve for Example X.4

190

Day	Concurrent Jobs	Rating
1	3	0.80
2	3	0.80
3	3	0.80
4	3	0.80
5	7	1.33
6	7	1.33
7	7	1.33
8	6	1.61

Day	Concurrent Jobs	Rating
9	4	1.07
10	4	1.07
11	5	1.34
12	6	1.61
13	4	1.07
14	4	1.07
15	4	1.07
16	4	1.07

Day	Concurrent Jobs	Rating
17	4	1.07
18	4	1.07
19	4	1.07
20	3	0.80
21	4	1.07
22	4	1.07
23	4	1.07
24	3	0.80
25	4	1.07

Day	Concurrent Jobs	Rating
26	4	1.07
27	3	0.80
28	3	0.80
29	2	0.54
30	2	0.54
31	2	0.54
32	1	0.27
33	1	0.27
34	1	0.27

Figure 13.23 Congestion rating for job chart 1 of Figure 13.24

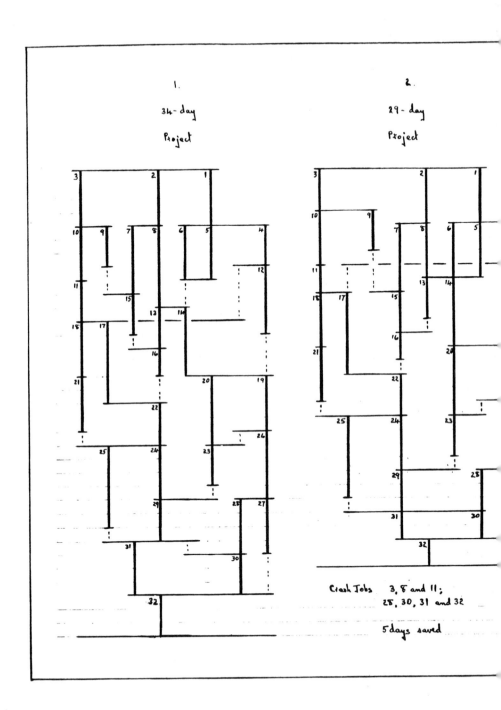

1.

34 - day

Project

2.

29 - day

Project

Crash Jobs   3, 8 and 11;
              28, 30, 31 and 32

5 days saved

Figure 13.24 Project cr

192

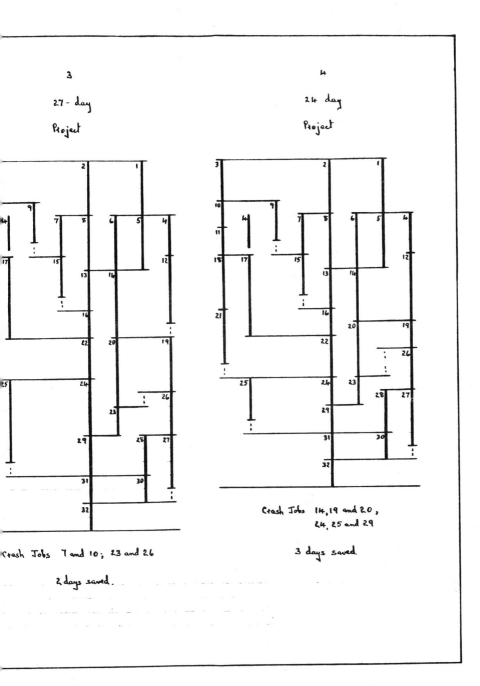

3

27- day

Project

4

24 day

Project

Crash Jobs 7 and 10; 23 and 26

2 days saved.

Crash Jobs 14, 19 and 20,
24, 25 and 29

3 days saved

indicated by Figure 13.21

Job Number	Job Sequence	Job Duration days	Manpower Requirement
1	0 : 1	1	2C.
2	0 : 2	1	2F.
3	0 : 3	1	2C.
4	0 : 4	3	3P.
5	1, 2, 3, 4 : 5	1	F. R.
6	1, 2, 3, 4 : 6	4	4F. W.
7	4 : 7	4	3F.
8	4 : 8	4	2C.
9	6 : 9	2	2F. R.
10	7, 8 : 10	6	4F.
11	6, 9 : 11	2	2F. W.
12	9, 11 : 12	2	2F.
13	10 : 13	6	4F.
14	7, 8, 10 : 14	4	2F.
15	12, 13 : 15	2	2E.
16	14 : 16	4	3F.
17	16 : 17	4	3F. R.
18	17 : 18	2	2F. R.
19	9 : 19	8	4F.
20	10 : 20	6	3P.
21	0 : 21	2	2F.
22	21 : 22	1	2F. R.
23	0 : 23	3	3P.
24	22 : 24	2	3F. W.
25	24 : 25	2	2F. R.
26	25 : 26	1	2F.
27	26 : 27	1	2F.
28	23, 26 : 28	1	2F. R.
29	28 : 29	2	2F.
30	29 : 30	2	2F.
31	27, 30 : 31	1	2E.
32	31 : 32	2	2E.
33	0 : 33	20	3E.
34	1 – 33 : 34	4	3F.
35	1 – 33 : 35	6	3F.

Figure 13.25   Job Sheet for Example X.5

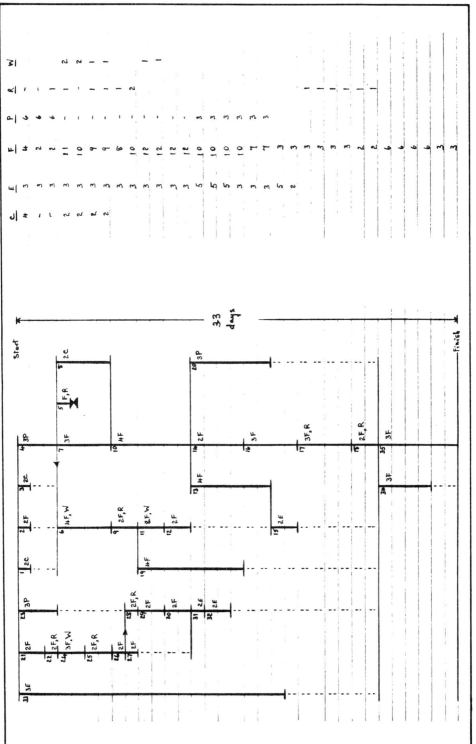

Figure 13.26 Job chart and resource array for Example X.5

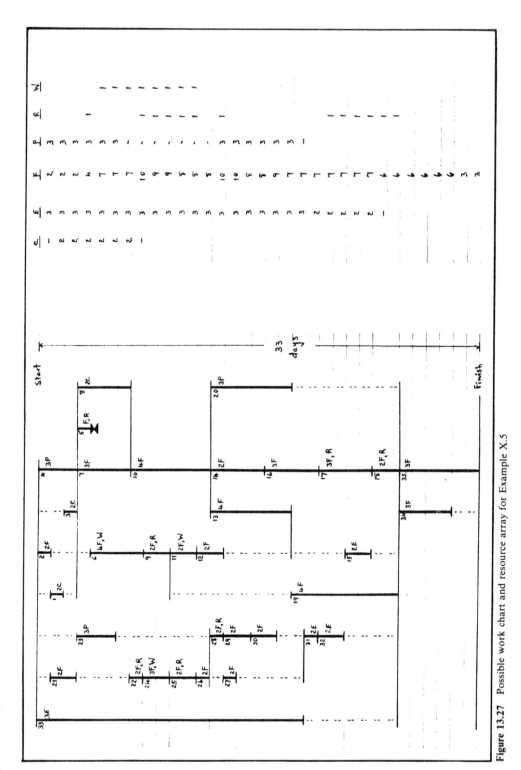

Figure 13.27 Possible work chart and resource array for Example X.5

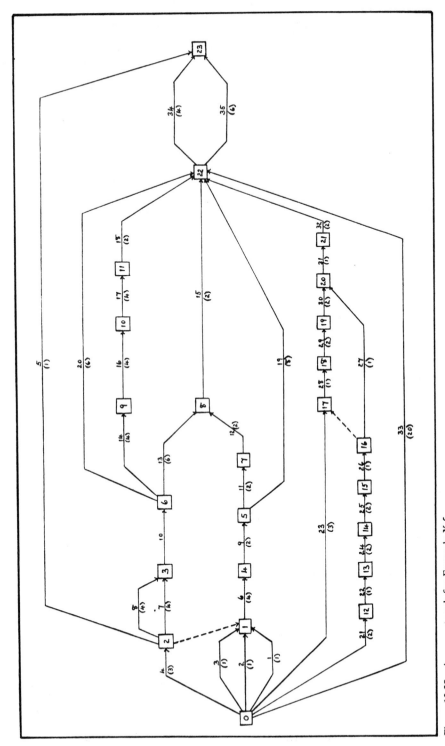

Figure 13.28 Arrow network for Example X.5

Job Number	Job Sequence	Job Duration days	Manpower Requirement
1	0 : 1	4	4F, 2R
2	0 : 2	6	4B
3	1, 2 : 3	6	3F
4	1, 2 : 4	4	3E
5	1, 2 : 5	3	4F, R
6	1, 2 : 6	2	3F
7	3, 4 : 7	4	2P
8	3, 4 : 8	2	2P
9	3, 4 : 9	4	4F
10	5 : 10	2	6F
11	6 : 11	8	2F, W
12	6 : 12	2	2F
13	8, 9, 10 : 13	2	2E
14	8, 9, 10 : 14	3	3F
15	7, 11, 12 : 15	2	4F
16	12 : 16	2	2F
17	12 : 17	7	4F
18	16 : 18	3	W
19	13 : 19	6	4F, 2R
20	13, 15 : 20	2	3F
21	13, 14 : 21	4	4F
22	20 : 22	8	4B
23	17, 18, 20 : 23	6	3F, W
24	19, 21 : 24	5	4F
25	22, 23 : 25	7	6F
26	24, 25 : 26	5	4F

Figure 13.29 Job sheet for Example X.6

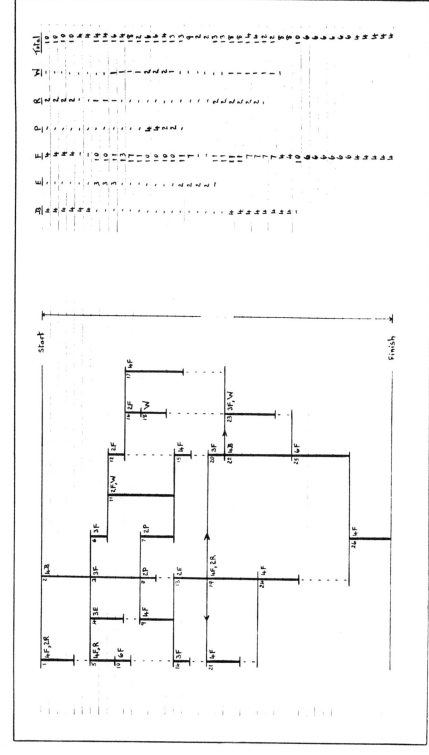

Figure 13.30 Job chart and resource array for Example X.6

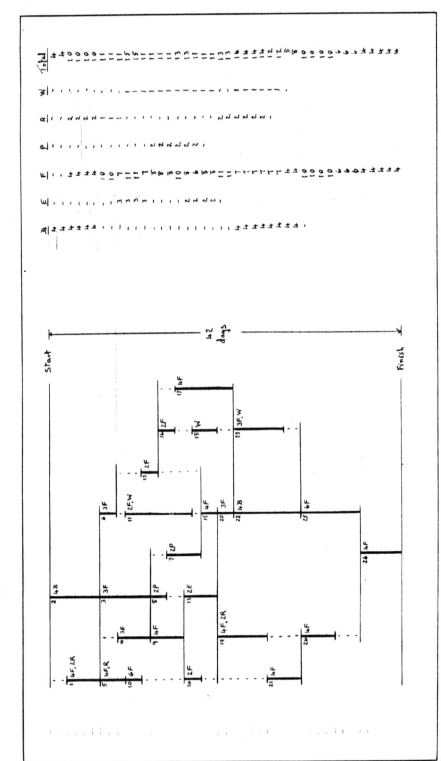

Figure 13.31 Work chart and resource array for Example X.6

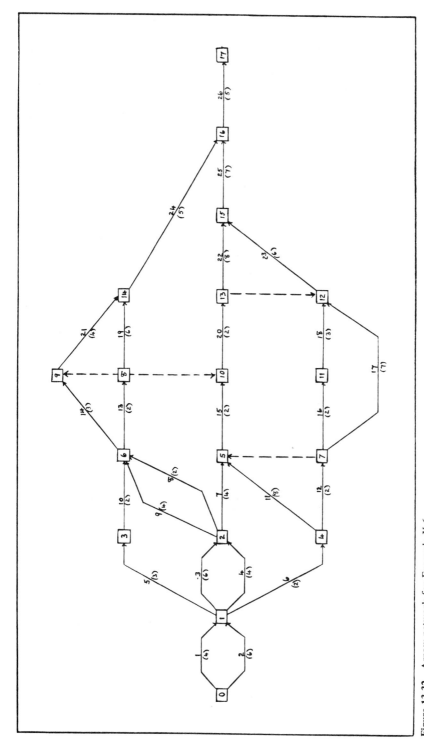

Figure 13.32 Arrow network for Example X.6

Job Number	Job Sequence	Standard Duration, days	Working Cost, £	Manpower Required	Crash Duration, days	Working Cost, £	Manpower Required
1	0 : 1	4	800	4F, W	2	1 200	8F, 2W
2	1 : 2	2	450	3F	1	570	6F
3	1 : 3	5	750	3F, R	3	910	6F, 2R
4	1 : 4	3	620	2E	2	720	4E
5	0 : 5	3	570	4B	2	650	6B
6	5 : 6	3	600	3F, W	3	600	3F, W
7	6 : 7	4	830	6F	2	1 100	12F
8	5 : 8	4	850	4P	2	1 050	8P
9	5 : 9	6	1 750	3F	3	2 050	6F
10	0 : 10	10	2 030	4C	5	2 650	8C
11	10 : 11	3	470	2F, W	2	550	3F, 2W
12	8, 11 : 12	3	520	2E	2	610	4E
13	10 : 13	5	800	2P	3	950	4P
14	13 : 14	3	490	2F, W	2	560	3F, 2W
15	2 : 15	4	820	3F	2	980	6F
16	2 : 16	5	1 100	4F	3	1 400	7F
17	3, 16 : 17	3	600	2F, W	2	700	3F, 2W
18	4, 9 : 18	3	580	2E	3	580	2E
19	4, 9 : 19	4	900	2F, R	2	1 300	4F, 2R
20	7, 19 : 20	6	1 350	2F	3	1 750	4F
21	15, 17, 18 : 21	3	820	2F, W	2	950	3F, 2W
22	20, 21 : 22	5	1 240	4F	3	1 550	7F
23	12, 14 : 23	3	670	2E	2	760	3E
24	12, 14 : 24	4	800	2F	2	910	4F
25	22, 23, 24 : 25	6	1 300	2F	3	1 560	4F
		?	21 710	334 days	?	?	?

Figure 13.33 Example X.7: job sheet for Project X.7.1

Job Number	Job Sequence	Standard Working			Crash		
		Duration, days	Cost, £	Manpower Required	Duration, days	Working Cost, £	Manpower Required
1	0 : 1	3	560	3F	2	630	5F
2	1 : 2	3	600	4F, R	3	600	4F, R
3	1 : 3	4	830	3F, R	2	930	6F, 2R
4	0 : 4	5	870	4F	3	1120	7F
5	3, 4 : 5	3	840	2F, W	3	480	2F, W
6	5 : 6	6	1800	4F, R	3	2750	8F, 2R
7	6 : 7	3	710	2F	2	920	3F, 2W
8	3 : 8	3	580	4P	3	580	4P
9	8 : 9	4	690	2F, R	2	500	4F, 2R
10	7, 13, 9 : 10	4	720	3F	2	820	6F
11	0 : 11	6	1350	2C	3	3000	6C
12	2, 11 : 12	3	500	4F	3	500	4F
13	12 : 13	3	520	4P	2	750	6P
14	12 : 14	5	1050	3F	3	1350	5F
15	12 : 15	4	860	3F, W	2	1070	6F, 2W
16	2, 11 : 16	5	990	3C	3	1180	5C

(continued overleaf)

17	14,15,16:17	5	1100	3F,W	3	1400	5F
18	14,15,16:18	3	470	2P	2	580	3P
19	14,15,16:19	3	530	2P	2	640	3P
20	14,15,16:20	6	1250	4F	4	1490	6F
21	6,10:21	5	1050	3P	3	1300	5P
22	6,10:22	3	520	2F,W	2	828	3F,2W
23	6,10:23	6	1450	4F,R	3	950	8F,2R
24	21,22,23:24	4	920	3F,W	2	1140	6F,2W
25	21,22,23:25	9	1600	4F	4	2050	6F
26	17,18,23:26	3	750	2P	2	950	3P
27	17,18,23:27	4	850	3F,W	2	1200	6F,2W
28	19,20:28	5	1120	2F,R	3	1550	4F,2R
29	19,20:29	6	1650	4F	3	970	8F
30	24,25:30	3	720	2F	3	720	2F
31	26,27:31	4	900	4F,R	2	1150	8F,2R
32	28,29:32	4	860	3F	2	1060	6F
33	30,31,32:33	4	840	3F	2	1060	6F
		?	29 690	490 days	?	?	?

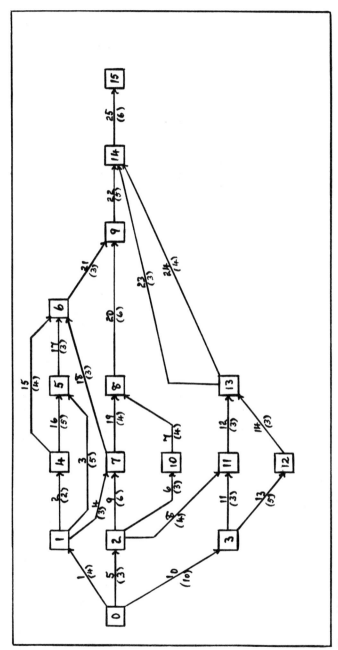

Figure 13.35 Arrow network for Project X.7.1

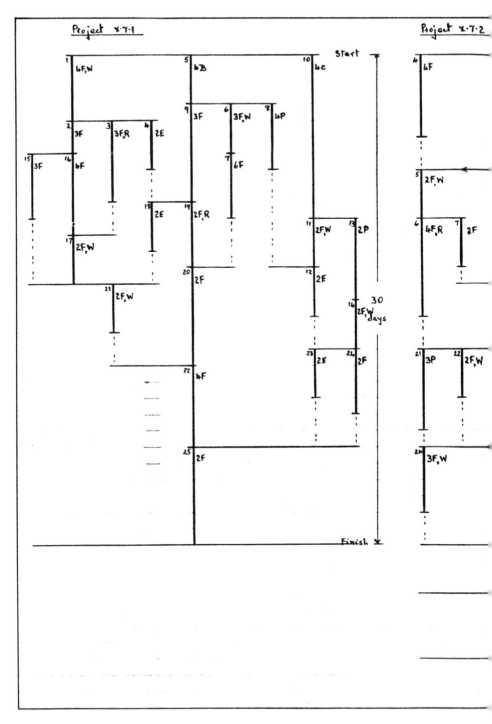

Figure 13.36 Ex job chart

Manpower Array

Day	B	C	E	F	P	R	W	Total
1	4	7	–	11	–	–	1	23
2	4	7	–	11	–	–	1	23
3	4	7	–	11	–	–	1	23
4	–	7	–	21	4	2	2	36
5		7	2	23	4	3	1	40
6		7	2	19	4	3	1	36
7		7	2	26	4	2	–	41
8		7	–	25	4	1	1	38
9		7	–	25	4	1	1	38
10		7	2	23	8	1	2	43
11		3	2	22	6	3	2	38
12		–	2	20	6	3	3	34
13		–		20	2	3	3	28
14			2	13	2	2	1	20
15			2	18	6	1	2	29
16			2	20	4	1	3	30
17			–	16	4	–	3	23
18			–	14	–	–	2	16
19			2	17	3	1	2	25
20			2	16	3	1	1	23
21			2	18	3	2	1	26
22			–	16	3	2	–	21
23				14	3	2	–	19
24				14	–	2	–	16
25				18	2	1	2	23
26				16	2	–	2	20
27				15	2	–	2	19
28				15	–	–	2	17
29				13	1	–	–	14
30				13	1			14
31				6	1			7
32				6	1			7
33				2	–			2
34				3				3
35				3				3
36				3				3
37				3				3
Total	12	73	24	549	83	41	42	824

Start — 37 days — Finish

Network nodes: 11 3C, 16 3C, 14 3F, 15 3F,W, 17 3F,W, 18 2P, 19 2P, 20 4F, 28 2F,R, 29 4F, 26 2P, 27 3F,W, 32 3F, 31 4F,R, 33 3F

power array for Example X.7

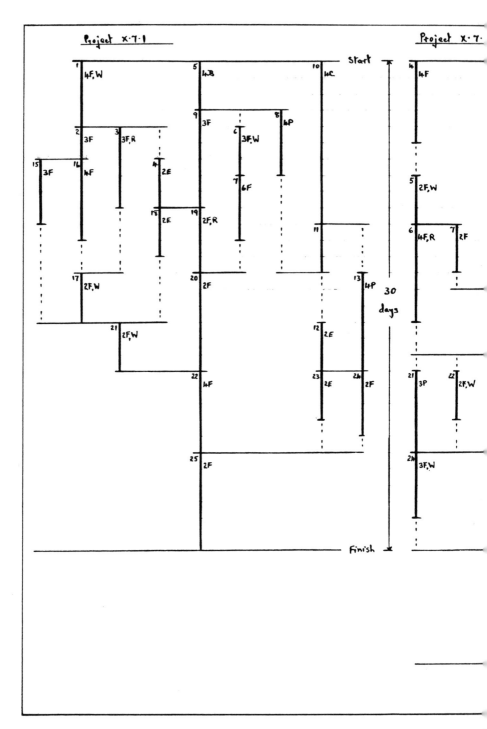

Figure 13.37 Work chart

Start

37 days

Finish

Diagram labels (network/bar chart, left): 3C; 16 3C; 3F; 15 3F,W; 18 2P; 19 2P; 20 4F; 7 3F,W; 28 2F,R; 29 4F; 6 2P; 27 3F,W; 31 4F,R; 32 3F; 3 3F

Day	B	C	E	F	P	R	W	Total
1	4	7	-	11	-	-	1	23
2	4	7	-	11	-	-	1	23
3	4	7	-	11	-	-	1	23
4	-	7	-	18	4	2	1	32
5		7	-	23	4	3	1	35
6		7	-	19	4	3	1	34
7		7	2	23	4	2	1	39
8		7	2	25	4	1	1	40
9		7	2	25	4	1	1	40
10		7	2	20	4	1	1	35
11		3	2	28	4	3	2	42
12		-	2	18	4	3	2	29
13			-	18	4	3	2	27
14			-	16	4	2	2	24
15			-	15	4	1	1	21
16			-	15	4	1	1	21
17			2	13	4	-	2	21
18			2	13	4	-	2	21
19			2	14	4	1	2	23
20			2	19	3	1	2	27
21			2	15	3	1	2	23
22			2	15	3	1	2	23
23			-	13	3	1	1	18
24				17	3	2	1	23
25				18	2	1	2	23
26				18	2	1	2	23
27				18	2	1	2	23
28				18	-	1	2	21
29				14		1	-	15
30				13		1		14
31				9		1		10
32				9		1		10
33				5		-		5
34				3				3
35				3				3
36				3				3
37				3				3
Total	12	73	24	549	85	41	42	826

ower array for Example X.7

Crash Jobs	Days Saved	Cost	Cheapness Rating
5	1	£ 80	80
25	3	260	86 2/3
22 and 24	2	420	210
13, 19 and 21	2	650	325
1, 2, 3, 7, 9, 10, 11, 14, 16, 17 and 20	6	2 820	470
Project	14	4 230	302

Figure 13.38 Project cost-duration tabulation for Project X.7.1

210

Crash Jobs	Days Saved	Cost	Cheapness Rating
3	2	£ 100	50
33	2	220	110
10 and 32	2	300	150
19, 20, 21 and 23	3	1290	430
25, 28, 29 and 31	2	1450	725
6, 9, 13 and 14	1	1590	1590
1, 5 and 11	1	1970	1970
Project	13	6920	532

Figure 13.39 Project cost-duration tabulation for Project X.7.2

211

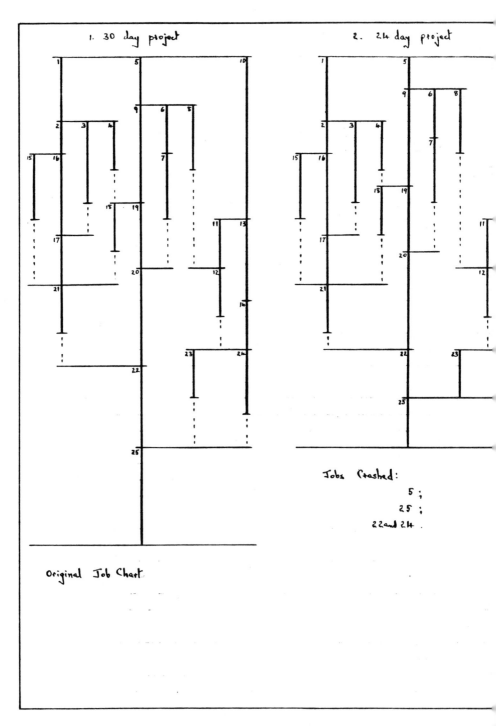

Figure 13.40 Project crash

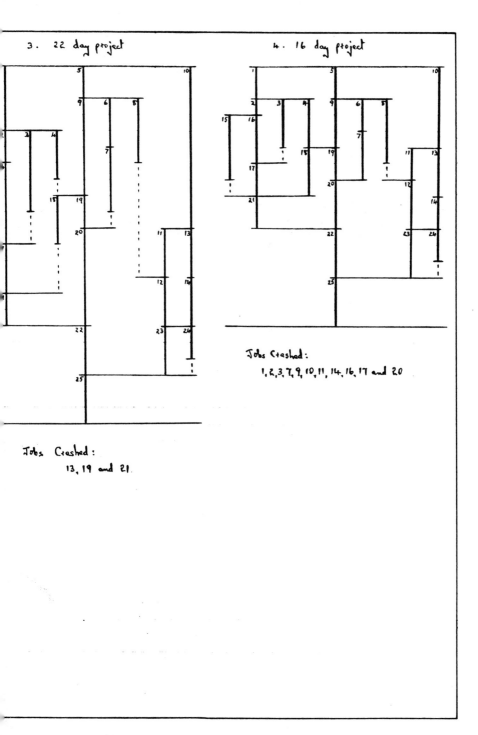

3.   22 day project

4.   16 day project

Jobs Crashed:
1, 2, 3, 7, 9, 10, 11, 14, 16, 17 and 20

Jobs Crashed:
13, 19 and 21.

as indicated by Figure 13.38

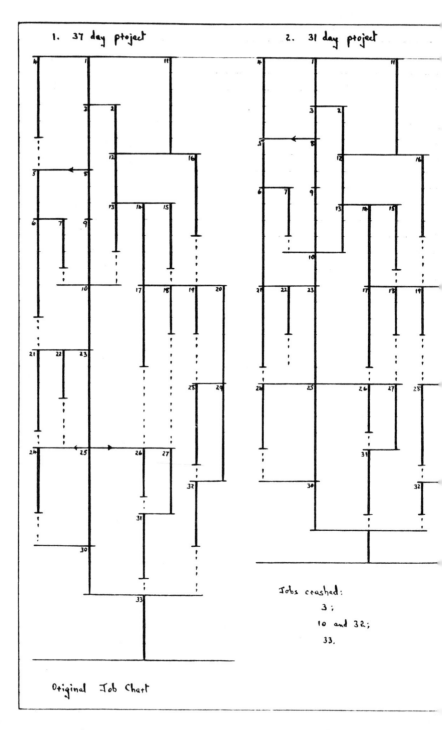

1. 37 day project 2. 31 day project

Original Job Chart

Jobs crashed:
3;
10 and 32;
33.

Figure 13.41 Project crash

214

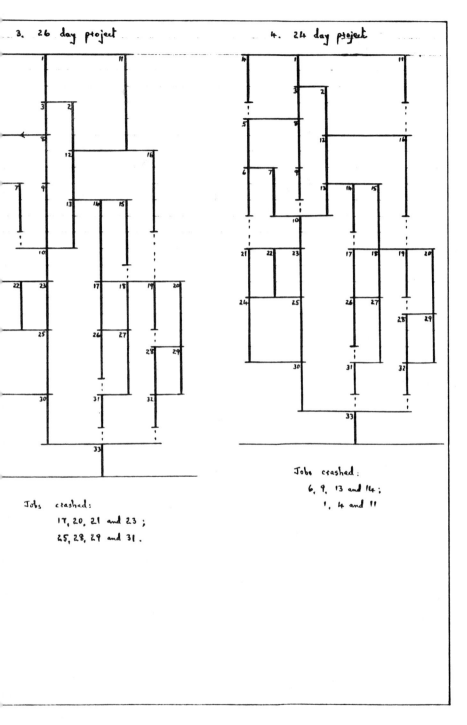

Jobs crashed:

17, 20, 21 and 23 ;
25, 28, 29 and 31 .

Jobs crashed:

6, 9, 13 and 14;
1, 4 and 11

as indicated by Figure 13.39

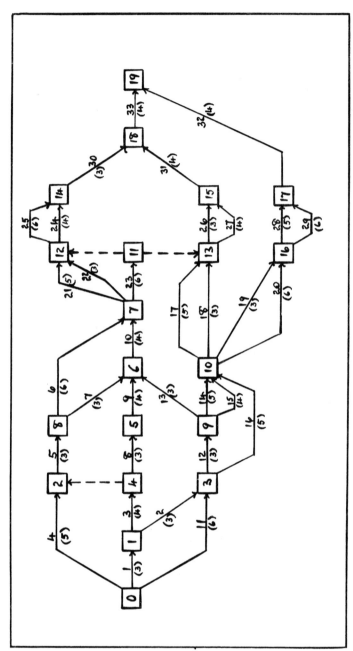

Figure 13.42 Arrow network for Project X.7.2

Figure 13.43 Job sheet for Project X.8.1

Job Number	Job Sequence	Standard Working Duration days	Standard Working Cost, £	Standard Working Manpower	Crash Working Duration days	Crash Working Cost, £	Crash Working Manpower
1	0 : 1	5	1 250	3F, W	3	1 850	5F, 2W
2	0 : 2	3	700	2F	2	800	3F
3	1 : 3	3	720	2F, R	3	720	2F, R
4	1 : 4	4	980	3F	2	1 580	6F
5	1 : 5	6	1 470	2F, W	4	1 870	3F, 2W
6	0 : 6	3	690	4P	2	910	6P
7	6 : 7	3	750	2P, R	2	970	3P, 2R
8	7 : 8	4	1 010	3P, R	3	1 260	4P, 2R
9	6 : 9	5	1 230	2P	3	1 700	4P
10	0 : 10	6	1 700	4C	3	2 900	8C
11	0 : 11	4	1 000	3E	2	1 650	6E
12	11 : 12	6	1 640	2E	3	2 190	4E
13	11 : 13	4	1 100	2E	2	1 710	4E
14	13 : 14	5	1 220	3E	3	1 820	5E
15	3, 8 : 15	4	960	3F, 2W	2	1 960	6F, 4W
16	3, 8 : 16	5	1 190	3F, W	3	1 800	5F, 2W
17	4, 9 : 17	4	1 050	3F	2	1 650	6F
18	9, 10 : 18	4	970	3P	2	1 570	6P

(continued overleaf)

	2	2		1	2			
19	4F	700	3	4F	700	3	12,18:19	
20	4F,2R	1580	2	2F,R	980	4	15:20	
21	3F,2W	1000	2	2F,W	720	3	15:21	
22	3F	740	3	3F	740	3	5,16,17:22	
23	8P	1980	2	4P	990	4	5,16,17:23	
24	7C	1650	3	4C	1050	5	14,19:24	
25	6F	950	2	4F	710	3	20,21:25	
26	3F	820	2	2F	740	3	20,21:26	
27	7F,4W	2350	3	4F,2W	1250	5	22:27	
28	3F	790	3	3F	790	3	22:28	
29	8C	1620	2	4C	820	4	22:29	
30	4F	1400	2	3F	780	4	28,29:30	
31	6C	1550	2	3C	930	4	23,24,26,27:31	
32	6E	2160	4	4E	1660	6	23,24,26,27:32	
33	7F	1920	3	4F	1300	5	25:33	
34	3F	650	3	3F	650	3	33:34	
35	5P	1700	3	3P	1200	5	33:35	
36	6F,4W	1900	2	3F,2W	880	4	33:36	
37	5F	1480	3	3F	1080	5	34,35,36:37	
38	6F	1500	2	3F	960	4	30,31,32:38	

218

Job Number	Job Sequence	Standard Working Duration days	Cost £	Manpower	Crash Working Duration days	Cost £	Manpower
51	0 : 51	11	3 550	3F	6	4 950	6F
52	51 : 52	3	770	2F, W	2	980	3F, 2W
53	52, 58 : 53	4	970	2E	2	1 350	4E
54	51 : 54	5	1 170	3P	5	1 170	3P
55	51 : 55	4	1 150	2F, R	2	1 750	4F, 2R
56	0 : 56	8	2 740	2F, R	4	4 000	4F, 2R
57	56 : 57	8	4 680	2F, W	6	5 250	3F, 2W
58	56 : 58	4	1 200	2F	2	1 600	4F
59	55 : 59	6	1 750	2F, R	3	2 650	4F, 2R
60	59 : 60	4	880	3F	2	1 480	6F
61	55 : 61	8	2 960	2F	4	3 760	4F
62	14, 53, 54, 57 : 62	3	650	3E	2	840	5E
63	14, 53, 54, 57 : 63	4	900	3F	4	900	3F
64	14, 53, 54, 57 : 64	6	1 460	3C	3	2 260	6C
65	62, 63 : 65	4	960	3F, W	2	1 770	6F, 2W
66	62, 63 : 66	7	2 210	2F	4	2 820	4F
67	24, 60, 61, 65 : 67	5	1 430	3F	3	1 830	5F
68	24, 60, 61, 65 : 68	4	1 050	3E	4	1 050	3E
69	24, 60, 61, 65 : 69	5	1 380	2F	3	1 750	4F
70	64, 66 : 70	4	1 100	4F, 2R	2	2 320	8F, 4R
71	70 : 71	3	720	2F, W	2	1 010	3F, 2W
72	70 : 72	5	950	4F, 2R	3	1 950	7F, 4R
73	68 : 73	4	800	3E	2	1 400	6E
74	121, 69, 73 : 74	5	900	3F	3	1 300	5F
75	71, 72, 74 : 75	3	740	3F	2	940	5F
	Project	?	37 070	384	?	?	?

Figure 13.44 Job sheet for Project X.8.2

Job Number	Job Sequence	Standard Duration, days	Standard Working Cost, £	Standard Manpower	Crash Duration, days	Crash Working Cost, £	Crash Manpower
101	:101	6	1 960	2F	3	2 560	4F
102	:102	2	2 840	2F, W	2	2 840	2F, W
103	:103	4	3 200	2F	2	3 600	4F
104	:104	9	3 750	3F	4	4 100	5F
105	:105	3	840	2E	3	840	2E
106	:106	9	1 520	3F	3	2 410	6F
107	:107	4	980	2F, W	2	1 590	4F, 2W
108	:108	9	1 440	4F, 2R	3	3 050	8F, 4R
109	:109	2	3 700	2F, R	2	3 700	2F, R
110	:110	4	880	2C	2	1 280	4C
111	:111	7	1 410	3C	4	2 310	6C
112	:112	3	420	E	3	420	E
113	:113	5	650	F	3	850	2F

Job No.	Qty	Value	Code	Job No.	Qty	Value	Code
116	5	1 250	2F, W	:116	3	1 850	4F, 2W
117	5	1 100	2F	:117	3	1 550	4F
118	4	1 000	2F, R	:118	2	1 600	4F, 2R
119	5	1 280	3F, R	:119	3	1 900	5F, 2R
120	3	720	2E	:120	2	840	3E
121	5	1 940	2F, W	:121	3	2 580	4F, 2W
122	7	1 820	2F	:122	4	2 420	4F
123	3	1 050	3C	:123	3	1 050	3C
124	6	2 100	3F	:124	4	2 500	5F
125	8	1 900	2C	:125	4	2 700	4C
126	4	1 300	3P	:126	2	1 900	6P
127	5	1 400	2F	:127	3	1 800	4F
128	6	1 990	3P	:128	3	2 890	6P
Project	?	44 360	352	?	?	?	?

Figure 13.45   Job sheet for Project X.8.3

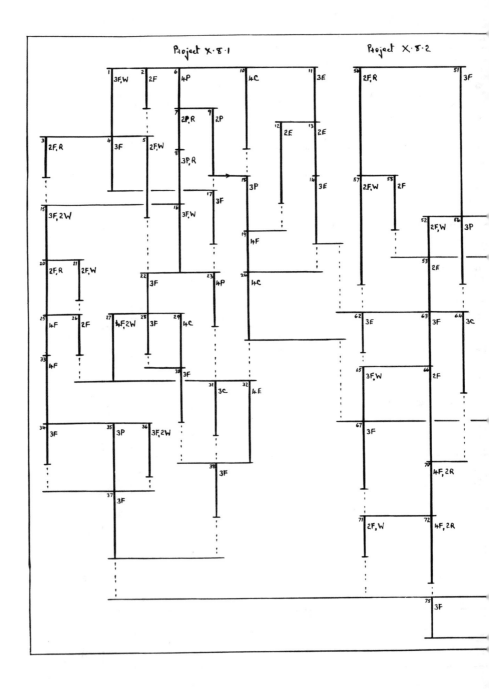

Figure 13.46 Job chart

222

Project X.8.3

Day	C	E	F	P	R	W	Total
1	4	3	18	4	3	2	34
2	4	3	18	4	3	2	34
3	4	3	18	4	3	2	34
4	4	3	16	4	4	2	33
5	4	4	14	4	4	1	31
6	4	4	18	4	5	1	36
7	2	4	21	5	4	2	38
8	2	4	21	5	4	2	38
9	2	5	17	6	1	2	33
10	2	5	17	6	1	2	33
11	3	5	24	3	-	5	40
12	3	5	23	6	1	5	43
13	3	6	25	3	2	5	44
14	3	1	22	3	2	5	36
15	3	3	21	3	3	3	36
16	7	2	17	7	2	2	37
17	7	2	16	7	2	2	36
18	4	2	18	7	4	1	36
19	11	3	29	7	3	3	56
20	11	3	29	-	3	3	49
21	7	3	29	-	3	3	45
22	7	2	19	-	1	2	31
23	3	2	21	-	-	3	29
24	6	6	15	-	-	1	28
25	6	4	19	-	-	2	31
26	6	4	16	-	-	2	28
27	6	7	17	3	-	3	36
28	2	7	20	3	-	3	35
29	2	7	20	3	-	3	35
30	2	3	20	3	2	2	32
31	2	3	17	3	2	-	27
32	2	3	13	3	2	-	23
33	2	3	13	3	2	-	23
34	2	3	9	6	2	1	23
35	2	-	12	6	2	1	23
36	-		14	3	2	1	20
37			9	3	2	-	14
38			9	3	2		14
39			5	3	-		8
40			5	-			5
41			3				3
42			3				3
	144	127	710	137	76	79	1 273

Node labels in the project network diagram:
2F | 2F.W | 4F,2R
2F | 3F | 2F.W | 2C | 2F.R
3F | 2F | 3C
2F.R | F | E
3P | 2F.W | 2F | 2F.R | 3F.R
2E
2F.W | 2F | 3C
3F | 2C
3P
3P
2F

ce array for Example X:8

223

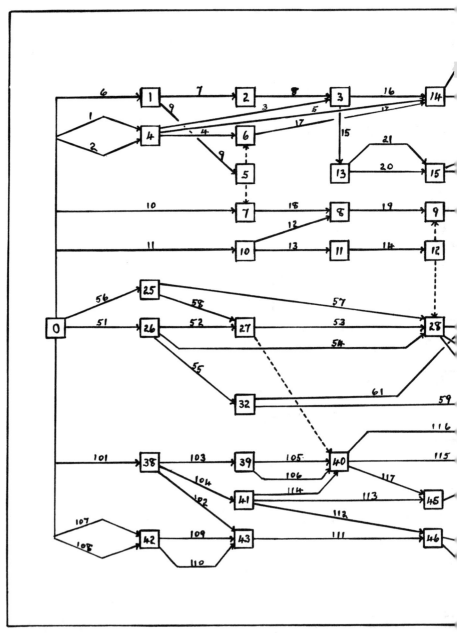

Figure 13.47 Arrow network for Example X.8

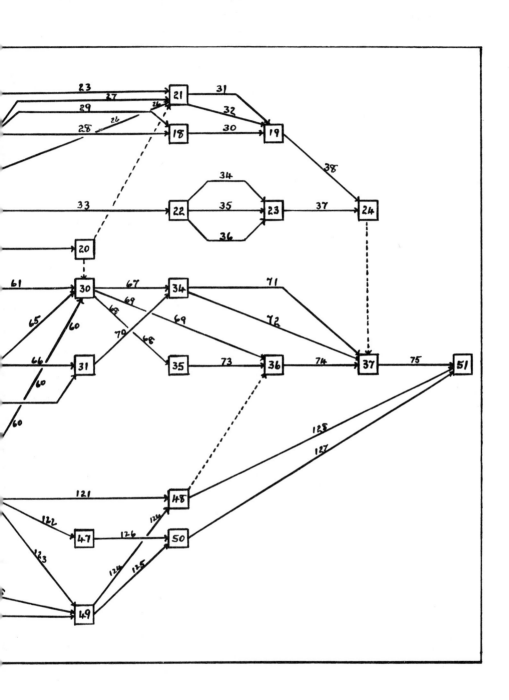

225

Figure 13.48 Job sheet for a larger project (1)

Job Number	Job Sequence	Standard Duration days	Working Cost £	Crash Duration days	Working Cost £
1	0 : 1	4	1 150	2	1 350
2	0 : 2	6	1 820	3	2 420
3	1,2 : 3	5	1 490	3	1 690
4	3 : 4	6	1 800	3	2 400
5	4 : 5	6	1 930	4	2 350
6	3,8 : 6	4	1 050	4	1 050
7	6 : 7	3	750	3	750
8	1,2 : 8	4	990	2	1 180
9	1,2 : 9	9	3 000	6	3 650
10	1,2 : 10	5	1 520	3	1 800
11	0 : 11	6	1 840	3	2 400
12	0 : 12	4	1 210	2	1 430
13	0 : 13	8	2 500	4	3 250
14	11,12,13 : 14	7	2 240	4	2 810
15	11,12,13 : 15	5	1 490	3	1 690
16	11,12,13 : 16	9	3 040	5	3 810
17	9,10,16 : 17	5	1 400	3	1 600
18	9,10,16 : 18	6	1 750	3	2 320
19	14,15,21 : 19	8	2 320	4	3 320
20	0 : 20	13	4 020	7	7 960
21	20 : 21	4	1 130	4	1 130
22	14,15,21 : 22	5	1 600	3	1 800
23	22 : 23	5	1 510	3	1 680
24	0 : 24	5	1 580	3	1 730
25	24 : 25	10	3 200	6	4 050
26	24 : 26	5	1 470	3	1 690
27	26 : 27	7	2 110	4	2 740
28	14,25,27 : 28	5	1 270	3	1 480
29	20,28 : 29	6	1 750	4	2 200
30	20,28 : 30	9	2 810	5	3 700
31	20,28 : 31	13	3 860	8	4 840
32	5, 7 : 32	7	2 040	4	2 600
33	17,18,19 : 33	10	3 100	5	4 200
34	17,18,19 : 34	6	1 730	3	2 370
35	19 : 35	4	1 150	4	1 150
36	23,35 : 36	6	1 690	3	2 300
37	32,33 : 37	15	5 020	8	6 400
38	32,33 : 38	5	1 390	3	1 610
39	32,33 : 39	9	2 840	5	3 700
40	38 : 40	7	2 130	4	2 720

(continued)

Job Number	Job Sequence	Standard Duration days	Working Cost £	Crash Duration days	Working Cost £
41	38:41	10	3150	5	4150
42	34,36:42	11	3720	6	4790
43	34,36:43	9	2270	5	3170
44	37,40:44	7	1770	4	2320
45	37,40:45	7	1820	4	2410
46	45:46	6	1490	4	1880
47	37,40:47	10	2550	5	3600
48	39,41,42,43:48	5	1300	5	1300
49	39,41,42,43:49	8	1990	4	2800
50	39,41,42,43:50	12	3150	6	4370
51	39,41,42,43:51	11	2800	6	3820
52	29,30,31:52	6	1540	3	2140
53	29,30,31:53	9	2300	5	3080
54	29,30,31:54	13	2990	8	4150
55	54,87,88:55	10	2510	6	3360
56	54,87,88:56	5	1300	3	1500
57	56:57	5	1220	3	1500
58	57:58	7	1800	4	2370
59	56:59	10	2500	5	3480
60	90,38,52,53:60	7	1790	4	2410
61	55,60:61	4	1010	4	1010
62	61:62	7	1810	4	2430
63	55,60:63	21	4500	14	6050
64	55,60:64	9	2260	5	3040
65	58,59:65	8	2100	4	3000
66	44,46:66	9	2200	5	3000
67	47,48,49:67	7	1710	5	2130
68	49,50,61:68	8	1980	4	2810
69	49,50,61:69	5	1250	5	1250
70	51,62:70	5	1240	3	1450
71	51,62:71	9	2270	5	3040
72	64,65:72	7	1750	4	2310
73	64,65:73	9	2190	5	2970
74	66,67,68,69:74	8	2090	4	2910
75	66,67,68,69:75	11	2400	6	3500
76	74,75:76	5	1200	5	1200
77	66,67,68,69:77	12	2460	6	3750
78	63,70,71,72,73:78	5	1140	3	1400
79	63,70,71,72,73:79	4	900	4	900
80	77,78,79:80	6	1310	3	1900

(continued overleaf)

Figure 13.48 *(continued)*

Job Number	Job Sequence	Standard Working Duration days	Standard Working Cost £	Crash Working Duration days	Crash Working Cost £
81	14, 25, 27 : 81	8	1 750	4	2 550
82	14, 25, 27 : 82	14	2 700	7	4 000
83	14, 25, 27 : 83	5	1 200	3	1 400
84	14, 25, 27 : 84	10	2 050	5	3 000
85	28, 83, 84 : 85	5	1 050	5	1 050
86	28, 83, 84 : 86	8	1 620	4	2 460
87	29, 81, 82 : 87	7	1 500	4	2 130
88	29, 81, 82 : 88	5	990	3	1 200
89	85, 86 : 89	5	1 100	5	1 100
90	89 : 90	4	830	2	1 050
91	90 : 91	6	1 310	3	1 880
92	85, 86 : 92	8	1 570	4	2 340
93	91, 92 : 93	4	850	4	850
94	91, 92 : 94	9	1 850	6	2 400
95	91, 92 : 95	5	1 060	3	1 260
96	95 : 96	13	2 420	8	3 380
97	95 : 97	6	1 130	3	1 700
98	91, 92 : 98	10	1 970	5	3 000
99	58, 59, 93 : 99	8	1 550	4	2 280
100	64, 94, 96, 99 : 100	7	1 390	4	1 950
101	64, 94, 96, 99 : 101	11	2 260	6	3 450
102	97 : 102	4	850	4	850
103	98, 102 : 103	9	1 930	6	2 500
104	98, 102 : 104	13	2 500	7	3 620
105	103, 104 : 105	5	1 000	3	1 200
106	79, 100, 101 : 106	6	1 150	3	1 700
107	76, 80, 105, 106 : 107	8	1 540	4	2 270
108	107 : 108	6	1 090	3	1 600
109	76, 80, 105, 106 : 109	5	950	5	950
110	109 : 110	7	1 400	4	2 020
111	109 : 111	10	1 990	5	3 000
112	76, 80, 105, 106 : 112	8	1 620	4	2 370
113	111, 112 : 113	7	1 500	4	2 050
114	111, 112 : 114	3	660	3	660
115	108, 110 : 115	3	650	2	800
116	108, 110 : 116	6	1 180	3	1 710
117	115, 116 : 117	3	570	3	570
118	113, 114 : 118	5	900	3	1 130
119	117, 118 : 119	5	870	3	1 050
	Project	?	216 950	?	?

Figure 13.49 Example X.10: job sheet for a larger project (2)

Job Number	Job Sequence	Standard Working Duration days	Standard Working Cost £	Crash Working Duration days	Crash Working Cost £
1	0 : 1	7	2 050	4	2 650
2	1 : 2	5	1 520	3	1 700
3	2 : 3	6	4 810	3	5 400
4	1 : 4	8	2 250	4	3 100
5	1 : 5	6	1 830	4	2 250
6	1 : 6	13	4 030	7	5 210
7	1 : 7	10	2 970	5	4 000
8	7 : 8	7	4 750	4	5 350
9	1 : 9	8	2 620	4	3 400
10	7,9 : 10	10	3 180	5	4 200
11	7,9 : 11	7	2 100	4	2 720
12	7,9 : 12	6	2 760	3	3 330
13	3, 4 : 13	7	2 090	4	2 810
14	3, 4 : 14	5	1 540	5	1 540
15	5, 6, 14 : 15	6	2 780	3	3 450
16	5, 6, 14 : 16	5	2 020	3	2 300
17	5, 6, 14 : 17	8	2 370	4	3 150
18	6, 11, 12, 21 : 18	8	2 510	5	3 220
19	18 : 19	8	2 820	4	3 650
20	19 : 20	11	3 550	6	4 610
21	0 : 21	22	7 240	11	9 440
22	6, 11, 12, 21 : 22	5	1 630	3	1 800
23	6, 11, 12, 21 : 23	3	1 050	3	1 050
24	23 : 24	3	990	3	990
25	24 : 25	4	1 410	2	1 580
26	0 : 26	40	12 000	20	16 050
27	13, 15, 16 : 27	5	1 580	3	1 760
28	27 : 28	6	1 740	3	2 470
29	13, 15, 16 : 29	9	4 650	5	5 520
30	13, 15, 16 : 30	6	4 800	3	5 390
31	29, 30 : 31	5	1 430	3	1 640
32	30 : 32	9	2 640	5	3 450
33	28, 31 : 33	7	2 220	4	2 810
34	28, 31 : 34	4	1 250	4	1 250
35	8, 10, 17 : 35	6	1 770	3	2 390
36	30, 35 : 36	5	1 490	3	1 730
37	8, 10, 17 : 37	8	2 430	4	3 160
38	32, 36, 37 : 38	5	1 610	3	1 840
39	32, 36, 37 : 39	7	2 270	4	2 880
40	32, 36, 37 : 40	4	1 190	4	1 190

(continued overleaf)

Figure 13.49 *(continued)*

Job Number	Job Sequence	Standard Duration days	Working Cost £	Crash Duration days	Working Cost £
41	19, 40 : 41	7	2 450	4	3 030
42	19, 40 : 42	12	4 200	6	5 150
43	25, 26, 30 : 43	8	2 390	4	3 090
44	25, 26, 30 : 44	5	1 310	3	1 530
45	25, 26, 30 : 45	10	2 650	5	3 700
46	25, 26, 30 : 46	7	1 740	4	2 310
47	25, 26, 30 : 47	9	2 300	5	3 130
48	25, 26, 30 : 48	7	1 810	4	2 380
49	43, 44 : 49	5	1 100	5	1 100
50	20, 43, 44 : 50	5	1 330	3	1 510
51	20, 43, 44 : 51	5	1 270	3	1 460
52	45, 46, 47, 48 : 52	5	1 270	3	1 490
53	45, 46, 47, 48 : 53	6	1 490	3	2 040
54	45, 46, 47, 48 : 54	4	980	2	1 170
55	45, 46, 47, 48 : 55	4	1 050	4	1 050
56	54, 55 : 56	4	1 080	2	1 280
57	53 : 57	4	1 020	2	1 200
58	49, 51 : 58	7	1 790	4	2 340
59	33, 34, 38 : 59	4	1 110	4	1 110
60	50, 39, 41, 42, 59 : 60	7	1 770	7	1 770
61	60 : 61	5	1 270	3	1 450
62	61 : 62	4	1 000	2	1 260
63	50, 39, 41, 42, 59 : 63	5	1 220	3	1 400
64	50, 39, 41, 42, 59 : 64	9	2 300	6	2 880
65	61 : 65	5	1 250	3	1 420
66	63, 64 : 66	4	990	2	1 210
67	50, 39, 41, 42, 59 : 67	7	1 700	4	2 250
68	50, 39, 41, 42, 59 : 68	10	2 460	5	3 390
69	50, 39, 41, 42, 59 : 69	8	1 980	4	2 550
70	67, 68, 69 : 70	5	1 240	5	1 240
71	52, 56, 57, 58 : 71	8	2 440	4	3 200
72	71 : 72	6	1 550	3	2 150
73	71 : 73	7	1 720	4	2 230
74	52, 56, 57, 58 : 74	17	5 500	9	7 090
75	52, 56, 57, 58 : 75	4	960	2	1 140
76	75 : 76	5	1 230	5	1 230
77	75 : 77	8	2 100	4	2 880
78	75 : 78	4	1 040	2	1 200
79	78 : 79	6	1 560	3	2 150
80	78 : 80	4	1 020	4	1 020

(continued)

Job Number	Job Sequence	Standard Working Duration days	Cost £	Crash Working Duration days	Cost £
81	75 : 81	9	1 350	5	2 140
82	75 : 82	7	1 200	4	1 750
83	75 : 83	5	1 060	3	1 300
84	80 : 84	6	1 130	3	1 770
85	62, 65, 66 : 85	5	1 000	5	1 000
86	62, 65, 66 : 86	5	970	3	1 140
87	71, 72, 73, 74, 85, 86 : 87	7	1 100	4	1 690
88	62, 65, 66 : 88	8	1 310	4	1 950
89	71, 72, 73, 74 : 89	4	820	2	1 020
90	71, 72, 73, 74 : 90	7	1 450	4	2 010
91	71, 72, 73, 74 : 91	6	1 300	3	1 930
92	76, 77, 79 : 92	11	2 090	6	3 080
93	84 : 93	11	1 950	6	2 920
94	84 : 94	8	1 810	4	2 660
95	81, 82, 83 : 95	9	1 870	5	2 660
96	81, 82, 83 : 96	6	1 150	3	1 780
97	89, 90, 91, 92 : 97	4	950	4	950
98	87, 88 : 98	6	1 200	3	1 740
99	87, 88 : 99	5	1 060	3	1 230
100	98, 99 : 100	5	1 040	5	1 040
101	89, 90, 91, 92 : 101	12	2 410	6	3 310
102	100, 101 : 102	5	1 020	3	1 240
103	93, 94, 97 : 103	6	1 380	3	1 870
104	93, 94, 97 : 104	7	1 430	4	2 000
105	95, 96 : 105	9	1 680	5	2 550
106	103, 104 : 106	9	1 720	6	2 230
107	102, 106 : 107	7	1 370	4	1 960
	Project	?	215 400	?	?

Figure 13.50 Example X.11: job sheet for a larger long-term project (3)

Job Number	Job Sequence	Standard Working Duration (weeks)	Cost £	Crash Working Duration (weeks)	Cost £
1	0 : 1	6	1 920	4	2 920
2	1 : 2	6	3 040	3	4 480
3	1 : 3	5	2 790	3	3 730
4	3 : 4	6	2 360	3	3 890
5	3 : 5	4	1 170	4	1 170
6	5 : 6	5	1 500	3	2 520
7	3 : 7	6	3 110	4	4 230
8	7 : 8	8	3 220	4	5 320
9	1 : 9	8	3 750	5	5 210
10	1 : 10	9	3 680	5	5 600
11	7, 9, 10,28 : 11	11	4 620	7	6 580
12	7, 9, 10,28 : 12	6	2 750	3	4 260
13	1, 4 : 13	6	3 020	4	4 080
14	1, 4 : 14	8	3 690	4	5 730
15	6, 13, 14 : 15	20	3 770	12	8 350
16	6, 13, 14 : 16	7	2 440	4	3 960
17	6, 13, 14 : 17	10	3 330	5	5 870
18	8 : 18	6	3 010	3	4 610
19	11,16,17, 18 : 19	7	3 840	4	5 320
20	19 : 20	5	2 620	5	2 620
21	11, 16,17, 18 : 21	10	5 100	6	7 310
22	11,16,17,18 : 22	5	2 460	3	3 480
23	22 : 23	8	3 180	4	5 070
24	28 : 24	5	1 990	3	2 990
25	0 : 25	52	40 000	26	80 000
26	15, 25 : 26	5	4 860	5	4 860
27	0 : 27	7	4 900	4	6 440
28	27 : 28	6	3 520	4	4 510
29	27 : 29	5	3 040	3	4 540
30	28 : 30	7	3 470	4	5 270
31	29 : 31	6	3 810	3	5 010
32	31 : 32	3	1 700	3	1 700
33	29 : 33	8	2 630	4	4 600
34	29 : 34	6	2 470	3	3 900
35	33, 34 : 35	6	2 460	4	3 450
36	35 : 36	4	1 650	4	1 650
37	11, 36 : 37	5	2 590	3	4 620
38	33, 34 : 38	8	3 300	4	5 340
39	38 : 39	4	1 680	2	2 770
40	29 : 40	10	4 960	6	7 060

(continued)

Job Number	Job Sequence	Standard Duration (weeks)	Working Cost £	Crash Duration (weeks)	Working Cost £
41	40,59 : 41	7	3 420	4	5 240
42	41 : 42	8	4 210	4	6 210
43	6, 12, 30, 32 : 43	27	6 860	18	10 150
44	6, 12, 30, 32 : 44	7	2 270	4	3 690
45	6, 12, 30, 32 : 45	10	3 980	5	6 400
46	20, 21, 23, 24, 43 : 46	8	2 430	4	4 480
47	20, 21, 23, 24, 43 : 47	6	2 520	4	3 550
48	37, 39, 44, 45 : 48	25	6 950	15	11 170
49	37, 39, 44, 45 : 49	5	3 050	5	3 050
50	49 : 50	7	2 940	4	4 440
51	42, 49 : 51	5	5 100	5	5 100
52	42, 49 : 52	7	2 930	4	4 390
53	50, 51 : 53	9	3 670	6	5 170
54	50, 51 : 54	12	6 000	6	9 150
55	52 : 55	7	3 340	4	4 800
56	52 : 56	5	2 860	3	3 920
57	0 : 57	4	3 930	5	5 800
58	57 : 58	5	2 400	3	3 260
59	58 : 59	3	900	3	900
60	57 : 60	6	2 160	3	3 690
61	60 : 61	5	2 220	3	3 800
62	34, 61 : 62	7	3 180	4	3 990
63	40, 59 : 63	3	1 350	3	1 350
64	62, 63, 66 : 64	5	2 920	3	3 910
65	57 : 65	8	3 970	4	5 980
66	65 : 66	6	3 110	3	4 580
67	62, 63, 66 : 67	7	2 840	4	4 220
68	0 : 68	78	7 250	52	10 000
69	42, 64, 67 : 69	8	2 810	4	4 830
70	42, 64, 67 : 70	6	2 590	4	3 560
71	70 : 71	5	2 190	3	3 220
72	42, 64, 71 : 72	11	4 390	7	6 540
73	72 : 73	7	2 200	4	3 680
74	73 : 74	19	5 250	10	7 950
75	56, 69, 71 : 75	18	7 260	12	9 000
76	56, 69, 71 : 76	10	4 130	5	6 150
77	56, 69, 71 : 77	14	4 080	8	6 760
78	76, 77 : 78	4	1 450	4	1 450
79	26, 46, 47, 48 : 79	6	2 240	3	3 700
80	79 : 80	6	2 270	4	3 250

(continued overleaf)

Figure 13.50 *(continued)*

Job Number	Job Sequence	Standard Duration (weeks)	Working Cost £	Crash Duration (weeks)	Working Cost £
81	79 : 81	10	4 840	5	7 310
82	79 : 82	8	3 610	5	5 160
83	81, 82 : 83	7	3 150	4	4 690
84	79 : 84	6	2 190	3	3 570
85	79 : 85	6	2 220	4	3 280
86	75, 84, 85 : 86	8	3 160	4	5 300
87	86, 88, 89 : 87	6	3 050	3	4 550
88	79 : 88	12	4 760	6	7 920
89	79 : 89	8	3 230	4	5 200
90	79 : 90	15	6 660	10	10 000
91	86, 88, 89 : 91	7	3 420	4	5 030
92	86, 88, 89 : 92	10	4 280	6	6 260
93	53, 54, 55 : 93	14	5 770	10	8 010
94	93 : 94	11	5 150	6	7 510
95	93 : 95	7	2 480	4	4 010
96	93 : 96	6	3 200	6	3 200
97	80, 81, 82, 90, 94, 95, 96 : 97	20	9 210	10	18 420
98	80, 81, 82, 90, 94, 95, 96 : 98	10	3 740	6	5 700
99	74, 75, 78 : 99	6	2 760	3	4 220
100	99 : 100	10	3 990	5	5 560
101	74, 75, 78 : 101	10	4 630	6	6 650
102	74, 75, 78 : 102	8	2 940	4	5 040
103	74, 75, 78 : 103	11	3 880	6	5 310
104	68, 102, 103 : 104	5	1 750	5	1 750
105	100, 101, 104 : 105	8	3 600	4	7 200
106	100, 101, 104 : 106	5	2 620	3	3 630
107	105, 106 : 107	7	3 130	4	4 680
108	100, 101, 104 : 108	23	9 760	15	14 000
109	100, 101, 104 : 109	5	1 750	5	1 750
110	109 : 110	6	2 270	3	3 770
111	110 : 111	6	3 010	3	4 520
112	109 : 112	9	3 850	5	5 790
113	83, 87, 91, 92 : 113	6	2 740	4	3 700
114	83, 87, 91, 92 : 114	8	3 670	4	5 740
115	113, 114 : 115	4	1 310	4	1 310
116	83, 87, 91, 92 : 116	4	1 720	4	1 720
117	116 : 117	5	2 170	3	3 160
118	116 : 118	7	4 060	4	5 550
119	116 : 119	7	3 710	4	5 210
120	110, 115, 117, 118, 119 : 120	6	3 500	3	4 960

(continued)

Job Number	Job Sequence	Standard Duration (weeks)	Working Cost £	Crash Duration (weeks)	Working Cost £
121	110, 115, 117, 118, 119 : 121	10	5 250	6	7 210
122	120, 121 : 122	4	1 380	4	1 380
123	110, 115, 117, 118, 119 : 123	13	4 990	8	7 500
124	110, 115, 117, 118, 119 : 124	6	2 040	3	3 550
125	124 : 125	5	1 730	3	3 230
126	124 : 126	8	2 630	4	4 700
127	98, 107, 113, 114 : 127	5	2 010	3	3 000
128	127 : 128	5	2 040	5	2 040
129	123, 127 : 129	8	3 380	4	5 390
130	123, 125, 126, 128, 129 : 130	4	1 760	4	1 760
131	98, 107, 113, 114 : 131	5	2 250	3	3 340
132	108, 131 : 132	7	3 460	5	4 450
133	98, 107, 111, 112, 113, 114 : 133	17	7 200	10	10 310
134	122, 130, 132, 133 : 134	6	2 770	3	4 270
135	134 : 135	9	3 830	6	5 330
136	134 : 136	6	3 050	6	3 050
137	134 : 137	6	3 180	4	4 220
138	134 : 138	13	5 000	8	7 500
139	134 : 139	8	3 310	4	5 350
140	122, 130, 132, 133 : 140	8	3 170	4	5 250
141	122, 130, 132, 133 : 141	6	2 820	4	4 100
142	140, 141 : 142	5	1 990	3	3 000
143	140, 141 : 143	8	3 530	4	5 550
144	140, 141 : 144	6	3 000	6	3 000
145	140, 141 : 145	8	2 690	4	4 610
146	144, 145 : 146	5	2 490	3	3 570
147	135, 136, 137 : 147	7	2 170	4	3 630
148	135, 136, 137 : 148	5	2 060	5	2 060
149	148 : 149	6	2 710	3	4 240
150	146, 147, 149 : 150	6	3 050	4	4 550
151	138, 139, 142, 143 : 151	17	6 040	10	10 900
152	138, 139, 142, 143 : 152	5	2 550	3	4 000
153	146, 152 : 153	6	2 750	3	4 400
	Project	?	548 240	?	?

Figure 13.51 Example X.12: job sheet for a larger project (4)

Job Number	Job Sequence	Standard Working Duration (days)	Standard Working Cost £	Crash Working Duration (days)	Crash Working Cost £
1	0 : 1	8	310	4	510
2	1 : 2	12	1190	6	1470
3	1 : 3	6	260	4	370
4	3 : 4	9	440	6	600
5	4 : 5	4	210	4	210
6	3 : 6	5	360	3	440
7	3 : 7	8	480	4	700
8	1 : 8	9	370	5	590
9	1 : 9	12	600	7	920
10	2,5,6,7,8,9,22:10	4	280	2	390
11	10 : 11	8	320	4	500
12	10:12	6	360	3	530
13	10:13	9	630	5	860
14	2,5,6,7,8,9,22:14	7	510	4	660
15	2,5,6,7,8,9,22:15	10	620	5	980
16	15:16	7	290	4	410
17	2,5,6,7,8,9,22:17	17	850	10	1300
18	0 : 18	4	170	4	170
19	18:19	6	330	3	460
20	19:20	5	420	3	590
21	19:21	8	660	4	920
22	20,21,34:22	6	390	4	530
23	20,21,34:23	8	610	4	840
24	19:24	6	350	3	520
25	19:25	10	820	6	1080
26	24,25:26	10	770	5	1100
27	22,23,26,30:27	23	940	14	1450
28	22,23,26,30:28	6	290	6	290
29	28:29	10	500	5	740
30	24,25:30	4	290	4	290
31	24,25,34,35:31	7	480	4	660
32	24,25,34,35:32	9	610	5	910
33	0 :33	7	430	4	750
34	33:34	6	340	3	500
35	33:35	8	430	4	630
36	33:36	11	1010	6	1300
37	30,31,32:37	7	500	4	640
38	14,37,49:38	6	380	3	540
39	14,37,49:39	5	440	5	440
40	39:40	6	420	4	540

(continued)

Job Number	Job Sequence	Standard Duration (days)	Working Cost £	Crash Duration (days)	Working Cost £
41	39:41	6	330	3	470
42	30,31,32:42	15	610	10	870
43	0:43	9	440	5	640
44	43:44	5	350	5	350
45	48:45	10	630	6	850
46	43:46	8	320	4	500
47	36,44,45,46:47	9	360	5	550
48	30,47:48	8	450	4	690
49	30,47:49	6	450	4	650
50	14,49:50	6	420	3	560
51	14,49:51	4	150	4	150
52	14,49:52	7	360	4	510
53	42,48,50,51,52:53	4	310	4	310
54	42,48,50,51,52:54	6	370	3	520
55	42,48,50,51,52:55	9	400	6	560
56	42,48,60,51,52:56	10	730	6	950
57	42,48,50,51,52:57	8	650	4	830
58	11,12,13,14,16,39:58	20	1 240	10	1 750
59	11,12,13,14,16,39:59	9	480	5	680
60	17,59:60	5	350	3	450
61	17,59:61	7	520	4	650
62	17,59:62	8	570	4	750
63	17,59:63	6	410	6	410
64	27,29:64	9	540	5	720
65	27,29:65	8	420	5	590
66	38,40,41,53,54:66	15	780	10	950
67	38,40,41,53,54:67	6	330	4	500
68	29,55,56,57,67:68	5	270	5	270
69	29,55,56,57,67:69	9	550	5	740
70	63,64:70	5	350	3	520
71	63,69:71	8	420	5	560
72	29,55,56,57,67:72	6	390	4	550
73	29,55,56,57,67:73	4	360	4	360
74	29,55,56,57,67:74	6	320	3	450
75	72,73,74:75	8	680	4	910
76	68,70:76	6	470	3	630
77	71,75:77	7	360	4	530
78	66,60,61,62,63,64,65:78	5	300	5	300
79	66,60,61,62,63,64,65:79	12	920	7	1 200
80	66,60,61,62,63,64,65:80	8	560	4	760

(continued overleaf)

Figure 13.51 *(continued)*

Job Number	Job Sequence	Standard Duration (days)	Working Cost £	Crash Duration (days)	Working Cost £
81	66, 60, 61, 62, 63, 64, 65 : 81	5	380	3	450
82	58, 78 : 82	5	290	3	370
83	58, 78 : 83	7	310	4	450
84	58, 78 : 84	5	240	5	240
85	70, 80, 81 : 85	4	260	2	380
86	70, 80, 81 : 86	5	370	3	460
87	70, 80, 81 : 87	5	350	3	450
88	70, 80, 81 : 88	5	290	5	290
89	66, 60, 61, 62, 63, 64, 65 : 89	7	450	4	660
90	71, 75 : 90	27	1 300	18	1 500
91	76, 77, 79, 82, 83, 84, 85, 86, 87, 88, 89 : 91	5	240	3	420
92	76, 77, 79, 82, 83, 84, 85, 86, 87, 88, 89 : 92	9	360	6	520
93	91, 92 : 93	11	580	6	830
94	91, 92 : 94	7	320	4	460
95	94, 99, 100, 117 : 95	9	770	5	950
96	94, 99, 100, 117 : 96	5	260	5	260
97	96 : 97	6	430	4	550
98	76, 77, 79, 82, 83, 84, 85, 86, 87, 88, 89 : 98	5	340	3	430
99	98 : 99	6	410	3	550
100	98 : 100	4	200	4	200
101	98 : 101	9	360	6	520
102	98 : 102	6	300	3	440
103	94, 99, 100, 117 : 103	6	270	4	380
104	101, 102, 103, 118 : 104	8	360	4	570
105	101, 102, 103, 118 : 105	5	260	3	350
106	93, 95, 97, 104 : 106	10	490	6	690
107	105 : 107	7	490	4	650
108	105 : 108	9	640	5	860
109	105 : 109	7	450	4	610
110	105 : 110	7	520	4	680
111	76, 77, 79, 82, 83, 84, 85, 86, 87, 88, 89 : 111	7	560	5	680
112	111 : 112	7	460	4	600
113	111 : 113	8	320	4	540
114	112 : 114	6	300	6	300
115	112 : 115	8	410	4	630
116	113, 115, 118 : 116	7	370	4	510
117	76, 77, 79, 82, 83, 84, 85, 86, 87, 88, 89 : 117	10	730	6	950
118	117 : 118	7	490	4	660
119	117 : 119	8	490	5	490
120	117 : 120	5	330	3	450

(continued)

Job Number	Job Sequence	Standard Working Duration (days)	Cost £	Crash Working Duration (days)	Cost £
121	117 : 121	4	250	4	250
122	113 ,115 ,118 :122	9	540	5	740
123	116 : 123	7	490	4	630
124	109 ,110 ,123 :124	7	420	4	580
125	90 ,119 ,120 ,121 :125	13	930	8	1 200
126	90 ,119 ,120 ,121 :126	20	1 450	12	1 900
127	90 ,119 ,120 ,121 :127	5	370	3	460
128	90 ,119 ,120 ,121 :128	3	210	3	210
129	127 , 128 :129	10	590	6	800
130	127 , 128 :130	6	320	4	400
131	122 , 129 , 130 :131	3	190	3	190
132	93 , 95 , 97 , 104 :132	17	1 040	10	1 400
133	106 , 107 , 108 :133	10	810	6	1 200
134	124 , 132 , 133 :134	5	220	3	340
135	124 : 135	9	630	5	850
136	124 : 136	11	890	6	1 200
137	124 : 137	10	780	6	1 000
138	122 , 125 , 126 , 131 :138	12	960	6	1 300
139	124 , 132 , 133 , 138 :139	8	460	4	700
140	134 , 135 , 136 , 137 , 139 :140	5	250	5	250
141	134 , 135 , 136 , 137 , 139 :141	7	360	4	510
142	134 , 135 , 136 , 137 , 139 :142	9	380	5	570
143	134 , 135 , 136 , 137 , 139 :143	5	310	5	310
144	140 , 141 , 142 , 143 :144	9	350	6	480
145	140 , 141 , 142 , 143 :145	5	200	3	350
146	134 , 135 , 136 , 137 , 139 :146	6	260	3	420
147	134 , 135 , 136 , 137 , 139 :147	9	390	5	600
148	134 , 135 , 136 , 137 , 139 :148	5	230	3	400
149	146 , 147 , 148 :149	5	210	5	210
150	144 , 145 , 149 : 150	5	250	3	400
	Project	?	70 180	?	?

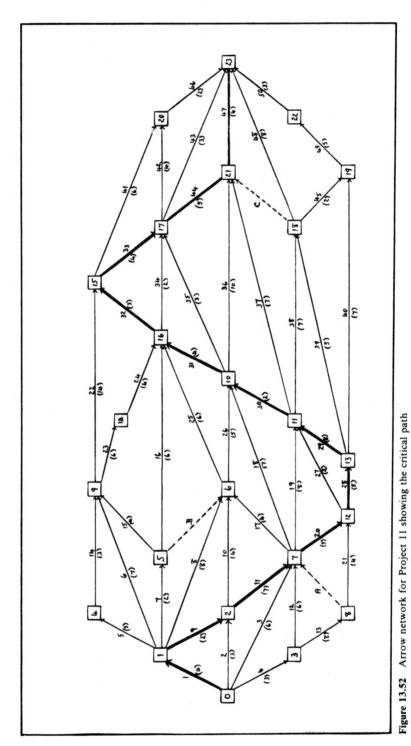

Figure 13.52 Arrow network for Project 11 showing the critical path

Uncertainty and Unreliability

The purpose of critical path analysis is to simplify project planning and scheduling by enabling supervisors to 'see' their way through a project before any work is started. The intention is to bring to light any difficulties that may arise and which would prevent the project being completed to plan, and to eliminate confusion from the minds of those concerned with planning and scheduling the project in its paper stage and with controlling the work when this begins. Very few of our planners have much knowledge of, or interest in, mathematical methods (a great pity, but that is another story), and simple non-mathematical methods have been developed so that a project can be planned and scheduled with the minimum of trouble.

It is regrettable that numerous attempts have been made to complicate critical path analysis, possibly to keep the techniques in the hands of the experts? It may be that these complications are responsible for the widespread refusal of maintenance and construction engineers to have anything to do with critical path analysis; either they tend to leave matters entirely in the hands of the 'black box experts' and so many plans go awry, or they claim that: 'We always do that sort of thing but never bother to write it down'; and so many plans still go awry.

Probability

Probably the worst of these complications is the introduction of the statistical concept of probability into the estimate of a job duration. This particular procedure requires three time estimates for each job, and the usual terminology and definitions of these estimates are

1   *The optimistic time, a,* which is the shortest possible time in which the job can be completed, on the assumption that all possible effort is brought to bear.

2   *The most likely time, M,* which is the time requirement of the job under normal standard conditions. This is the time estimate quoted as 'standard time' in the job sheet.
3   *The pessimistic time, b,* which is the longest time that the job is likely to require on the assumption that 'things' go wrong. This should be a time that is not likely to be exceeded more than once in a hundred repetitions of the job.

From these three time estimates various statistics can be calculated, and the statistics are used to calculate (so-called) probabilities that may be applied to each job and to the overall project to give, for each item considered, a range of durations within which it may be expected to get the job (or the project) completed. The statistics, the probabilities, and the ranges of durations are likely to include the total errors or inaccuracies made in developing the three time estimates.

Applying the Beta-distribution to the three time estimates, the expected duration, *d*, of any job is given by the relationship

$$d = (a + 4M + b)/6$$

and the use of this statistic can lead to some peculiar comparisons.

Consider, for instance, Job 12 in the job sheet of Figure 2.1 which has a standard time — the most likely duration — of 6 days. The crash time for this job is 4 days, and this must be taken as the most optimistic time since it is physically impossible to do the job in less time than this. If famine, pestilence, strikes, and any other major upheaval are excluded, the pessimistic time is 14 days. Estimation of this pessimistic time is not a straightforward business, and there is no specification list detailing the factors and causes to be admitted as possible (plausible?) and those to be ignored. Which factors can be admitted? The chain sling might break, and the part being hoisted would fall and very likely crack; this has never happened and care and maintenance of slings are undertaken to prevent such an occurrence, but it could happen sometime. Does one assume some rate of absenteeism? Does one allow an increased rate of absenteeism when a job is planned for week-end working? Is someone going to have an accident on the job? There are many such happenings which occur from time to time and which, when they occur, cause a long delay to the work.

Having sorted out all the possible causes of delay, and admitted those over which the planner has control — men, materials, tools and equipment, site, and working conditions — and obtained a figure of 14 days, then the expected duration of Job 12 is calculated to be

$$d = (4 + 24 + 14)/6 = 7 \text{ days}$$

which of course is 1 day more than the standard time of 6 days. Since experience and knowledge of everything involved in the job indicate that the job can be completed in 6 days, it is pertinent to ask where this extra day has come from. Has the job lost a day to statistics? Or have statistics imposed some uncertainty where none existed previously? Is the job to be considered a 6-day job or a 7-day job or what? This uncertainty principle can be thrown throughout the project, and using the estimate of variance:

$$s^2 = [(a - b)/6]^2$$

and assuming that the individual job variances are additive, joint estimates of uncertainty can be calculated for sequences of jobs and for the project overall.

Now another statistic is needed to transform variances (or the standard deviation derived from them) into probabilities, and that all-purpose warrior, Student's t, is thrown in. The use of Student's t with any calculated standard deviation permits the uncertainty factor to be related to a distribution of probabilities. Returning to Job 12, with the values quoted above, the standard deviation of uncertainty is

$$s = (a - b)/6 = (14 - 4)/6 = 1\tfrac{2}{3} \text{ days}$$

from which it may be deduced that, for a job scheduled to take 6 days and which has a minimum duration of 4 days, there is a probability of

0.68 that the job will be completed somewhere between $5\tfrac{1}{3}$ and $8\tfrac{2}{3}$ days $(d \pm s)$,

0.95 that the job will be completed somewhere between $3\tfrac{1}{3}$ and $10\tfrac{1}{3}$ days $(d \pm 2s)$

0.997 that the job will be completed somewhere between 2 and 12 days $(d \pm 3s)$

Apart from the fact that some of the values quoted in these probability statements are just impossible, the statements seem to admit a remarkable degree of laxity in the performance of a job. Can one really say that there is only 95 per cent probability that a 6-day job will be completed in $10\tfrac{1}{3}$ days? Do the probability statements mean anything? Should they be permitted to have any meaning?

Now consider this procedure applied to the critical jobs of the example project of Figure 2.1. It is simpler to consider the estimated and calculated values in tabular form and Figure A1.1 shows these values.

Job	optimistic time a	Standard time m	Pessimistic time b	Expected Duration d	Standard Deviation s	Variance s^2
3	4	6	14	7	$5/3$	$25/9$
5	2	4	8	$4\tfrac{1}{3}$	1	1
11	2	4	8	$4\tfrac{1}{3}$	1	1
13	4	5	12	6	$4/3$	$16/9$
15	5	7	15	8	$5/3$	$25/9$
Project	17	26	57 ?	$29\tfrac{2}{3}$		$84/9$

Figure A1.1   Uncertainty calculation for the project of Figure 2.1

It would be extremely unfortunate (for the project manager!) if all five jobs attained their pessimistic times. However, for the data tabulated for the five jobs in the critical path there is a standard deviation of uncertainty of

$$s = \sqrt{(84/9)} = 3 \text{ say (it is near enough for this discussion).}$$

If the relevant t-values are applied, ranges of project durations can be calculated as was done for Job 12. But with the overall project d-value of $29\tfrac{2}{3}$ days, the range of project durations for 0.997 probability is found to be

$$29\tfrac{2}{3} \pm (3 \times 3) \text{ or } 20\tfrac{2}{3} \text{ to } 38\tfrac{2}{3} \text{ days}$$

It seems that there is very little chance that the project will be completed in less than $20\tfrac{2}{3}$ days. In fact, this project was completed in the 17 days as planned and scheduled.

With regard to the quoting of three time estimates for any job, and the rest of the procedure discussed above, the following points must be borne in mind:

1 The 'most likely time' estimate is made as the result of an engineer's knowledge and experience of previous performances of the job or of similar work. He knows the capabilities of his men and of the equipment available. He knows the work that has to be done and the places where the men are to work; in many cases he will have done the job himself on some previous occasion. In most cases an 'optimistic' estimate or a 'pessimistic' estimate can vary between a fair assessment and a horribly wild guess. The guesses are likely to bear no relationship to any jobs done previously or likely to be done on the current or any future job.

2 As mentioned previously, it is not easy to decide what factors to admit for determination of the pessimistic time. Many flammable and explosive materials are used in the manufacture of chemicals, and fire and explosion are ever-present hazards; fortunately, they are infrequent, but they are not all that rare despite the precautions that are taken. Is pessimistic allowance made for a fire? Is additional labour allowed to guard against fire (this is done in some cases)? Regrettably, the death of a key workman can throw schedules completely haywire, and it is extremely difficult to include allowance for the death of a specified individual. Such occurrences have been reported recently and are likely to continue, but when and to whom is not likely to be determined beforehand. Quite often parts fail when equipment is under repair or is being installed. The likely incidence of such failures is known and is covered by having a stock of parts or made-up items available in case of need. The time allowance for pessimism in these cases is nil. Cases arise when something goes wrong after work has been started: a liner may be found cracked and in need of repair; baffle plates may be found to be corroded or broken; sludge in a reactor may be intractable; etc. There is no range of probabilities attaching to these occurrences; either they happen or they don't. History may give the incidence of previous occurrences and the delays caused by them will be known; any measures necessary to counter these delays will be planned in advance. Here again the time allowance for pessimism is nil.

3 Almost every job carried out can be completed in a shorter time than the standard. There is no question of probability in this; the shorter time is achieved only by premium expenditure. The only concern must be 'Is it worth incurring the additional expenditure?' No consideration should be given to a query such as 'can the job be completed in a shorter time and the premium expenditure charged against probability?' There is no probability of any sort that a job may be completed in less time than standard unless some additional facilities are made available.

4 Statements have been made to the effect that if three estimates of time are given for each job they are likely to be unbiased, whereas if one single estimate

for the standard, or most likely, time is required, this may be enlarged by the inclusion of some safety factor. The proponents of this procedure realised that it would be possible for a person who has given the three time estimates to conclude that all possibilities had been catered for so that an absolutely correct answer must be obtained. But such statements are untenable and must be refuted; if the authors of such statements have met a dishonest engineer or an engineer who did not know his job they have been very unlucky; they will have met many planners and engineers who do not understand black box methods and do not trust the results produced.

Where any work is to be done, the only time estimate for the job that is worth having is the honest opinion of the expert in charge of such work. Estimates of time are based on knowledge: the optimistic time for any activity is the shortest possible crash time for that activity and will be considered only if the crash time is worth saving; the pessimistic time for an activity cannot be known — the truly pessimistic statement is that the activity will never be completed, that is, it is of infinite duration, and in recent years there have been instances of large projects being abandoned or delayed indefinitely because of factors such as strikes and withdrawal of financial support which completely wreck any thought of planning and control by probabilistic procedures or by critical path procedures at all. One single estimate of standard time for a job has been proved to be sufficient, and a single estimate for crash time may be needed in some projects.

It is interesting to note that Battersby and Carruthers [*Operational Research Quarterly*, Vol. 17, pp. 359-80 (1966)] have commented that whereas it is often laborious to obtain single time estimates, the concept of three time estimates — even if it were valid and were to add significantly to the accuracy of the best estimate — is almost impossible to convey, let alone quantify. They add that where valid statistical distributions of activity times exist they differ widely from the Beta-distribution.

5   The main purpose of critical path analysis is to show which jobs are critical so that supervision may be concentrated on these jobs. Any difficulties likely to arise are discussed in the planning and scheduling stages, when necessary action may be taken or planned to ensure that time schedules and resource schedules are met and that floaters are permitted to float only within the amount of float available to them.

6   The author has neither experience nor knowledge of the application of probability techniques to a project carried out by men and supervisors unfamiliar with the job and with each other. He still begs leave to doubt whether probability does more than provide an escape clause for poor management.

Network complexity and project control

Once the job chart has been drawn and all the necessary calculations of cost functions and resource scheduling have been done and the final work chart agreed, the managers or supervisors still have to get the work done. The two important points on every project are the total cash to be spent on the project and the overall duration of the project; and it is not uncommon for the cash allocation to be exceeded and for the estimated overall duration to be very much less than actual. The 'overspend' and 'over-run' may be the result of lack of knowledge or lack of control. It is thought that if the lack of control is the result of insufficient supervision — major disturbances such as epidemics, strikes, subsidence, etc., which are not controllable by project management, are still excluded — then some forewarning of the likelihood of overspend or over-run may be given by the two following metrics

1   The *network complexity*, c, defined by

$$c = \text{(Number of jobs)/(Number of time lines between jobs)}$$

where the number of time lines includes the start and finish lines, and for an arrow network we would have

$$c = \text{(Number of activities)/(Number of events)}$$

2   The *network density, d*, defined by

$$d = \text{(Sum of job durations)/(Sum of job durations + Sum of floats)}$$

For the example project of Figure 2.1 with the charts of Figures 3.1 and 7.2 we have

Number of jobs   =   14
Number of time lines   = 9 (excluding any job end-line between a job and its float)
Sum of job durations   = 69 days (Figure 3.1)
    = 56 days (Figure 7.2)
Sum of job durations + floats =   88 days (Figure 3.1)
    =   56 days (Figure 7.2)

Figure A1.2 shows how the network density is increased by crashing the project to minimum duration and scheduling the resources.

Item	Job Chart Figure 3·1	Work Chart Figure 7·2
Number of Jobs	14	14
Number of Time-lines	9	9
Network Complexity c =	1·5	1·5
Sum of Job Durations	69	56
Sum of Job Durations plus Sum of Floats	88	56
Network Density d =	0·78	1·0

Figure A1.2   Comparison of network complexity and network density values

Day	Concurrent Jobs	Rating r	Day	Concurrent Jobs	Rating r	Day	Concurrent Jobs	Rating r
1	3	0.91	7	4	1.21	13	3	0.91
2	3	0.91	8	4	1.21	14	3	0.91
3	3	0.91	9	4	1.21	15	2	0.60
4	3	0.91	10	4	1.21	16	2	0.60
5	4	1.21	11	4	1.21	17	2	0.60
6	4	1.21	12	4	1.21			

Figure A1.3 Congestion ratings for the work chart of Figure 7.2

Note that crashing a project does not alter the network complexity, which has a theoretical minimum (for a one-job project) of 0.5; but crashing this project and scheduling the resources resulted in the maximum value of 1.0 for the network density, since all float had been eliminated. As either the network complexity or the network density increases, the likelihood of any disturbance causing an overspend or over-run also increases, and it has been postulated that control of large projects can be aided by frequent calculations of complexity and density on the remaining part of the project. True, if the values of c and d increase then the *chance* of any disturbance occurring has increased, but note from the work chart of Figure 7.2 that the project has a network density of 1.0 throughout, while the network complexity varies slightly through the project, falling to $2/3 = 0.6$ at the end of Day 14.

It is suggested that a more realistic measure of control metric is the number of jobs to be performed concurrently, and the value for each day can be compared with the average value for the project to give a *congestion rating, r*. This metric takes into account the number of jobs in the project, the sum of the job durations, and the overall duration of the project. For the example project, from the work chart of Figure 7.2 we have

Sum of job durations (which is the total job-days) = 56
Overall duration = 17
Average congestion = 3.3

and Figure A1.3 shows the congestion ratings for the work chart.

Obviously supervision is spread more thinly on days with a rating of 1.21 than on the other days, and a table such as that of Figure A1.3 shows which days are the more likely to be short on supervision so that additional supervision can be planned, if thought necessary, before work is started.

Glossary

Crash job A job whose duration can be reduced. Usually this involves extra effort — more men on the job or overtime working — and premium expenditure.

Critical job Any job on the critical path. The latest finish of one critical job must be the same as the earliest start of the next critical job.

Critical path That schedule which identifies the jobs that must be completed in succession, and determines the starting and finishing time of each of these jobs, so that the project may be completed without delay. Any other path is a 'slack' path.

Float The amount of time that the start of a job may be delayed without affecting any other job.

Floater Any job whose position in time can be moved without affecting a critical job. Floaters are found on slack paths.

Free float The additional (free or slack) time available for a job on the assumption that all jobs are started at their earliest possible start times. If a sequence of jobs forms a simple chain then the free float is considered to be attached to the last job in the chain.

Independent float The free (or slack) time available for a job irrespective of how any other jobs may be fixed in their own floats.

Job Any separate piece or item of work which marks a stage in the project or makes a definite advance towards completion of the project. It is necessary to include all restraints such as 'waiting for equipment from suppliers', 'waiting for furnace to cool', as jobs; they may require time even if they require no effort. But note that the actual definition of what constitutes a 'job' may depend on the project, the plan, or the schedule. In a small specialised project fitting a bolt to an inspection plate on the manlid of a reactor may constitute a job, while on another project installation of the complete reactor assembly may be listed as a single job.

Job chart An array of lines, each of which represents a job and whose length represents the job duration in some convenient scale.

Job line A line in the job chart which stands for a particular job in the project. Each job line is numbered to correspond with the number given to its job in the job sheet.

Job sheet A list of all the jobs and restraints in the project, with all relevant sequences, duration, and cost data for each job. Data may be included for standard working and for crash working.

Project The total work that has to be done from start to finish. Usually a project starts at the time a request is made for work to be carried out. It may include such items as 'engineering evaluations', 'chemical research', 'engineering design', 'management approval', which must be completed before any work can be started on site.

Time line A horizontal line in the job chart drawn at some particular point in time to relate the completion of two or more jobs to the start of another job or jobs.

Terms relating specifically to arrow networks

Activity An activity is a time-consuming element and represents a job or a restraint. It is indicated by a link of the network and is directed by an arrow. Activities are numbered.

Arrow network A network of lines each of which represents a job in the project or shows some special relationship between jobs. The lines are not drawn to scale and each line is labelled and directed.

Dummy A network link which shows only dependence of activities. A dummy has neither work nor time content, but is an essential part of a network when it is required to show a relationship between activities in different sequences.

Event A node of the network, that is a junction of two or more activities. Usually an event represents the start or finish of an activity. Events are numbered.

Bibliography

Astrachan, A., 'Better plans come from study of anatomy of an engineering job', *Busines 'Veek*, p60 (21 March 1959).

Battersby, A., Jetwork analysis', *Chemistry in Britain*, Vol. 2, pp 109-112 (1966).

Battersby, A. *Network Analysis*, Macmillan, London (1967).

Berge, C., *T ie Theory of Graphs and its Applications*, Methuen, London (1963).

Berman, P., 'Try critical path method to cut turnaround time 20%', *Petroleu n References*, p 65 (January 1962).

Buchan, J.R., and Luttrell, W.B., 'The critical path method of relocating departments', *Hospitals (Chicago), Vol. 43, No. 17, pp 79-82 (1969)*.

Buesnel, E.L., Hori, N.J., and Jurgens, R.C.M., *Unilever manual on network planning methods*, Unilever, London (1964).

Carruthers, H., and Battersby, A., *Advances in critical path methods* (1966).

Carson, J.M., 'Critical path project scheduling advancements provide greater control of plant expenditures by the process industries', The Lummus Co., 385 Madison Avenue, New York (1963).

Clarke, C.E., 'The optimum allocation of resources among the activities of a network', *Journal of Industrial Engineering*, Vol. 12, No. 1, p 11 (1961).

Clarke, C.E., 'The PERT model for the distribution of an activity time', *Journal of the Operational Research Society of America*, 10.3 (1962).

Duncan, J.B., and Russell, G.E., 'Critical path planning and scheduling', *R. and E. Report 61-214*, Monsanto, St Louis (1961).

Frazer, W., 'Progress reporting in the Special Projects Office', *Navy Management Review*, p 9 (April 1959).

Fletcher, A., and Clarke, G., *Management and Mathematics*, Chapter 4, Business Books (1972).

Ford, L.R. and Fulkerson, D.R., 'A simple algorithm for finding maximal network flows', *Canadian Journal of Mathematics*, Vol. 9, p 210 (1957).

Ford, L.R., and Fulkerson, D.R., *Flows in Networks*, Princeton University Press (1962).

Freeman, R., 'A generalised network approach to project activity sequencing', *IRE Transactions on Engineering Management*, EM-7 (September 1960).

Fulkerson, D.R., 'An out-of-kilter method for minimal cost flow problems', Paper P-1825, The Rand Corporation, Santa Monica, California (1955).

Fulkerson, D.R., 'A network flow computation for project cost curves', *Management Science*, Vol. 7, p 2 (January 1961).

Fulkerson, D.R., 'Expected critical path lengths in PERT networks', *Journal of the Operational Research Society of America*, Vol. 10, p 6 (1962).

Gass, S., and Saaty, T., 'The computational algorithm for the parametric objective function', *Nav. Res. Log Quart.*, Vol. 2, p 39 (1955).

Glaser, L.B., and Young, R.M., 'Critical path planning and scheduling — application to engineering and construction', *Chemical Engineering Process*, p 60 (November 1961).

Goode, H., and Machol, R., *System Engineering*, McGraw-Hill, New York (1957).

Goyal, S.K., 'Reduce project cost by de-crashing activities', *Journal of the Institute of Work Study Practitioners*, Vol. 15, No. 10, pp 645-646 (1971).

Goyal, S.K., (1972) 'Decrashing of activities in PERT networks, *ibid.*, Vol. 16, No. 9, pp 477-484 (1972).

Goyal, S.K., (1973), 'Application of work study to network scheduling calculations', *ibid.*, Vol. 17, No. 9, pp 634-636 (1973).

Grinyer, P.H., and Rogers, J.A., 'An investigation on the effectiveness of network analysis', *Work Study and Management Services*, pp 138-143 (April 1975).

Grubbs, F.E., 'Attempts to validate certain PERT statistics, or, Picking on PERT', *Journal of the Operational Research Society of America*, 10.6 (1962).

Hills, J., and Harris, M., 'Network analysis for new course planning', *Nursing Times*, pp 73-76 (20 May 1976).

Industrial Organisation, *A Bibliography of CPM and PERT*, Zurich (1963).

Kelley, J.E., 'Computers and operations research in road building', *Symposium Proceedings*, Case Institute of Technology (February 1957).

Kelley, J.E., *The Construction Scheduling Problem: Progress Report*, Univac Applications Research Centre, Philadelphia (April 1957).

Kelley, J.E., *Extension of the Construction Scheduling Problem: A Computationl algorithm*, Univac Applications Research Centre, Philadelphia (1958).

Kelley, J.E., 'Parametric programming and the primal dual algorithm', *Operations Research*, Vol. 7, p 3 (1959).

Kelley, J.E., *Critical Path Planning and Scheduling: Case Histories*, Maunchly Associates, USA (1960).

Kelley, J.E., 'Critical path planning and scheduling: mathematical basis', *Operational Research*, Vol. 9, p 296 (1961).

Kelley, J.E., and Walker, M.R., 'Critical path planning and scheduling', *Proc. Eastern Joint Computer Conference*, Boston, USA (1959).

Koopmans, T.C., *Activity Analysis of Production and Allocation*, Chapter 2, Monograph No. 13, Cowles Commission, New York (1962).

Leader Article, 'The analytical approach to work', *Financial Times* (11 February 1963).

Lowe, C.W., 'Critical path scheduling simplified', *Chemical Engineering*, p 170 (10 December 1962).

Lowe, C.W., 'Is PERT out-dated?', *Maintenance Engineering*, p 4 (16 January 1963).

Lowe, C.W., 'Job progress charts: the poor man's PERT', *The Manager*, p 57 (October 1963).

Malcolm, D.G., 'Extensions and applications of PERT as a system management tool', 7th National Conference, The Armed Forces Management Association, Washington, USA (1961).

Martino, R.L., 'New way to analyse and plan operations and projects will save you time and cash', *Oil/Gas World*, p 38 (September 1959).

Martino, R.L.; 'How critical path scheduling works', *Canadian Chemical Proceedings*, p 38 (February 1960).

Martino, R.L., (1963) 'Plain talk on critical path method', *Chemical Engineering*, p 221 (10 June 1963).

Mattozzi, M., and Lipinski, F., 'New approach to project scheduling: the control-operation technique', *ibid.*, p 135 (18 February 1963).

Moder, J.J., and Phillips, C.R., *Project Management with CPM and PERT*, Reinhold, New York (1964).

Mauchly, J.W., 'Critical path scheduling', *Chemical Engineering*, p 139 (16 April 1962).

Nakajima, S., 'PERT and other techniques for plant engineering and maintenance', International Plant and Maintenance Engineering Conference, London (17-21 June 1963).

Pickup, J.S., and Thwaites, D.A., 'Networks — a path to success', *The Manager* (December 1962).

Platts, C.V., and Wyant, T.G., 'Network analysis and the possibility of its use in education', *Education Review* (February 1969).

Pocock, J.W., 'PERT as an analytic aid for programme planning', *Journal of the Operational Research Society of America*, Vol. 10, No. 6 (1962).

Rogers, J.A., *The Extent and Effectiveness of Network Analysis*, PhD Thesis, The Graduate Business Centre, City University, London (1975).

Roots, W.K., 'Critical path network construction', *Works Management*, Vol. 20, No. 4, pp 6-9 (1967).

Sabezak, T.V., 'Network planning — a bibliography', *Journal of Industrial Engineering*, Vol. 13, No. 6 (1962).

Sage, B.W., 'Budgeting by network analysis', *Health and Social Service Journal*, pp 1020-1021 (5 May 1973).

Smith, M.E., *A Discussion of an Action-planning and Control technique*, International Minerals and Chemical Corp. (August 1960).

Special Projects Office, *PERT – Program Evaluation Research Task, Phase 1 Summary Report*, and *PERT – Program Evaluation Research Task, Phase 2 Summary Report*, Bureau of Ordnance, Dept. of the Navy, Washington, USA (1958).

Steinfeld, R.C., 'Critical path saves time and money', *Chemical Engineering* (28 November 1960).

Tabernacle, J.B., 'Network analysis', *Work Study*, pp 14-22 (March 1976).

UKAEA, *Network Analysis: a Programmed Text*, United Kingdom Atomic Energy Authority, London (1971).

White, D.J., Dinaldson, W.A., and Lawrie, N.L., *Operational Research Techniques, Vol. 1*, Business Books, London (1969).

Wood, A., and Wyant, T.G., 'Using network analysis on a learning sequence', *Industrial and Commercial Training*, Vol. 2, No. 12 (1970).

Woodgate, H.S., (1964) *Planning by Network*, Business Books, London (1964).

Index

256